AGRICULTURE ISSUES AND POLICIES

GRAPES: CULTIVATION, VARIETIES AND NUTRITIONAL USES

Agriculture Issues and Policies

Additional books in this series can be found on Nova's website under the Series tab.

Additional E-books in this series can be found on Nova's website under the E-book tab.

Food and Beverage Consumption and Health

Additional books in this series can be found on Nova's website under the Series tab.

Additional E-books in this series can be found on Nova's website under the E-book tab.

AGRICULTURE ISSUES AND POLICIES

GRAPES: CULTIVATION, VARIETIES AND NUTRITIONAL USES

RALPH P. MURPHY
AND
CHRISTOPHER K. STEIFLER
EDITORS

Nova Science Publishers, Inc.
New York

For permission to use material from this book please contact us:
Telephone 631-231-7269; Fax 631-231-8175
Web Site: http://www.novapublishers.com

LIBRARY OF CONGRESS CATALOGING-IN-PUBLICATION DATA

Grapes : cultivation, varieties and nutritional uses / editors: Ralph P. Murphy and Christopher K. Steifler.
p. cm.
Includes index.
ISBN 978-1-61470-950-3 (hardcover)
1. Grapes. 2. Grapes--Varieties. 3. Grapes--Utilization. I. Murphy, Ralph P. II. Steifler, Christopher K.
SB388.G73 2011
634.8--dc23 2011028900

Published by Nova Science Publishers, Inc. † New York

CONTENTS

PREFACE

In this book, the authors gather current research in the study of the cultivation, varieties and nutritional uses of grapes. Topics discussed in this compilation include characterization of wine industry residue and its application in foods; evaluation of grape seeds as a source of added value natural antioxidants; prohexadione-calcium as a regulator of vine growth; extraction and bioactive properties of grapeseed polyphenols; hypersensitivity and allergy to grapes; grapes as an alternative crop for water saving and reutilization of grape pomace as a solid medium of fermentation for the production of hydrolytic enzymes of industrial interest.

Chapter 1- Fullerenes, or buckminsterfullerenes in full, are molecules composed entirely of carbon, taking the form of a hollow sphere, ellipsoid, tube or ring. The C_n fullerene clasters consist of (n/2)-10 hexagonal and 12 pentagonal rings. Each carbon atom is bonded to three others and is sp^2 hybridised. There are several methods of the fullerenes synthesis: plasma-based method, resistive heating, corona discharge, combustion of carbon-containing substances. All of these methods are to the full extent described in monograph [1-4]. The light fullerenes molecules are very promising in the context of medical application, due to they demonstrate the following properties: antiHIV- protease activity, photodynamic DNA cleavage, free radical scavenger, antimicrobial action and use of fullerenes as diagnostic agents [1].

Chapter 2- Grape is the second most abundant fruit crop on the planet, following only by orange. According to the Food and Agriculture Organization (FAO), the world production of grapes was approximately 67 million tons. The world wine production reached nearly 27 million tons in 2008, indicating that 40% of the grape production was destined for wine production [1]. Approximately 20% of wine production is represented by pomace (peel and seed), with 5.4 million of tons of wine residues produced every year. These materials include biodegradable residue and their disposal creates serious environmental problems.

Chapter 3- Grape seeds are waste products of the winery and grape juice industry. These seeds contain lipid, protein, carbohydrates, and 5-8% polyphenols, depending on the variety. Grape seed polyphenols contain flavan-3-ols as monomers (catechin, epicatechin, gallocatechin, epigallocatechin and epicatechin 3-O-gallate) but also as procyanidin dimers, trimers, and highly polymerized procyanidins, aside from phenolic acid precursors (gallic acid). For this reason, the grape seed extract (GSE) is to be considered as a powerful antioxidant that prevents premature aging and disease (Schewe et al. 2001; Schewe et al. 2002).

Chapter 4- *Vitis vinifera* cv. Koshu grapes (Figure 1a) are widely cultivated in Central Japan, particularly Yamanashi Prefecture. Koshu grape is an indigenous grape cultivar that has been grown for more than a thousand years in Japan. Today, it is one of the most abundantly produced grape varieties in Japan. Koshu grape has been categorized in the cluster of oriental cultivars of *V. vinifera* (Figure 1b) (Goto-Yamamoto et al., 1998, 2006). The grapes are typically used in making Japanese high-quality wines. However, Koshu grapes have certain characteristics that differ from those of European grapes (*V. vinifera* L.) generally used for winemaking. For example, fresh berry weight of Koshu grapes at the time of harvest is more than twice those of Chardonnay, Sauvignon Blanc, Merlot, and Cabernet Sauvignon grapes, and Koshu grape skin becomes light purple during growth (Figure 1a). Moreover, Koshu grape juice has higher total phenolic (TP) content than juice from other white grape cultivars, and wine made from Koshu grapes also has higher TP content than white wine made from other European grape cultivars. Okamura and Watanabe (1981) have reported that the contents of caftaric and coutaric acids, which are the tartrate derivatives of caffeic and coumaric acids, respectively, are much higher in Koshu wines than in white wines made from Semillon, Chardonnay, and Riesling grapes. The high contents of these compounds are considered to affect the taste of Koshu wine (Yokotsuka, 1995). In addition to grape characteristics, the cultivation conditions of Koshu grapes differ from those of European grapes. In general, numerous Koshu grape vineyards demonstrate not guyot-style but shelf-style cultivation (overhead trellis) (Figure 1c).

Chapter 5- The production of premium quality grapes requires the control of several and multiple parameters in the vineyard. Therefore management of production yield is becoming increasingly more important for a quality vitiviniculture (Petrie and Clingeleffer, 2006). So far, manual cluster thinning is a widely used technique to reduce production in vigorous vines, producing an increasing in Colour Intensity (CI), Total Phenolic Index (TPI) and anthocyanins (Garcia-Escudero *et al.*, 2004). Although effective, manual cluster thinning is a very expensive operation because of the large labour requirements (Martinez de Toda *et al.*, 2003).

Chapter 6- Grapeseeds have taken attention from the scientific community because of the high content of galloylated proanthocyanidins. The percentage of galloylated moieties respect to the total is noted as galloylation degree, which highly affects the antioxidant and anticarcinogenic activities. The antioxidant properties of proanthocyanidins from grape pomace and grapeseed as affected by processing and extraction conditions, is discussed in this chapter. Thermal treatments and the maintaining of a reducing medium by bubbling nitrogen have been reported to increase the antioxidant activity. The high positive correlation between polyphenol content and antioxidant activity was demonstrated, thus being the selection of solvent a key factor for maximizing the antioxidant activity. Ethanol:water mixtures have been found to be the best solvent. Galloylated proanthocyanidins from white varieties have been reported to be more active than those anthocyanins from the red ones. The main differences between red and white varieties are also discussed in this chapter.

Chapter 7- The European Academy of Allergy and Clinical Immunology [1] proposed a mechanistic classification of food allergies:

Adverse food reactions are defined as any aberrant reaction after the ingestion of a food or food additive which may be the result of toxic or nontoxic food reactions.

*Toxic reaction*s can occur in anyone, provided a sufficient dose is ingested (eg, histamine in scombroid fish poisoning).

Nontoxic reactions depend on individual susceptibilities and may be the result of immune mechanisms (allergy or hypersensitivity) or nonimmune mechanisms (intolerance).

IgE-mediated food allergies have been most clearly delineated, but non-IgE-mediated immune reactions, especially of the gastrointestinal tract, are increasingly recognized.

Food intolerances probably account for the majority of adverse food reactions and may be caused by the food's pharmacologic properties (eg; headaches from tyramine in aged cheeses and jitteriness from caffeine in coffee or soft drinks), or by the host's unique susceptibilities, such as metabolic disorders (eg; lactase deficiency) or idiosyncratic responses.

Chapter 8- Water poor countries in Middle East and North Africa (MENA) region have to reallocate agricultural water for more sustainable use of scarce water resources. For a sustainable agriculture, governments should base their agricultural plans on water productivity, virtual water, and crop water requirement (CWR), acknowledging that these dimensions can play a vital role in a warmer climate and with a larger population.

Chapter 9- Grape is the most widely cultivated fruit crop in the world. From the world's total production of 60 million tonnes, about 68% of grapes are used for winemaking. Grape pomace is the main residue left after juice extraction from the grapes in the wine making industry and is formed from the skins, seeds and pieces of stem. [1] It constitutes about 16% of the original fruit. The average composition of this medium includes carbohydrates, fibre, fats, proteins and mineral salts. The main component of the fibre is lignin and then hemicelluloses, cellulose and pectin.

Chapter 10- Grapes and grape products are known worldwide and sustain a market with high added value, which uphold the demand for research in the fields of agronomy, technology and nutrition. This chapter focuses on giving a comprehensive view over the botanical and cultivation aspects of grapes; which varieties are suitable for technological application and the main consumed products; to clarify the influence of the bioactive compounds in the metabolic pathways that provides the nutritional benefits of grapes; and to update information on the important issue of waste generated in the grape industry.

In: Grapes
Editors: R. P. Murphy et al., pp. 1-49 ISBN 978-1-61470-950-3

Chapter 1

SOLUBILITY OF LIGHT FULLERENES AND FULLERENOL IN BIOCOMPATIBLE WITH HUMAN BEINGS SOLVENTS

Konstantin N. Semenov and Nikolay A. Charykov
St Petersburg State University, Russia

ABSTRACT

Experimental and literature data concerning the solubility of individual light fullerenes (C_{60} and C_{70}), industrial fullerene mixtures and water-soluble fullerenol in biocompatible with human beings binary (individual light fullerenes-oleic, linolic, linoleic acids; fullerenol-water) and multicomponent (fullerenes + natural fats and oils, essential oils) systems are presented and discussed, as well as application of investigated systems in medicine, pharmacology, food and cosmetic industry.

INTRODUCTION

Fullerenes, or buckminsterfullerenes in full, are molecules composed entirely of carbon, taking the form of a hollow sphere, ellipsoid, tube or ring. The C_n fullerene clasters consist of (n/2)-10 hexagonal and 12 pentagonal rings. Each carbon atom is bonded to three others and is sp^2 hybridised. There are several methods of the fullerenes synthesis: plasma-based method, resistive heating, corona discharge, combustion of carbon-containing substances. All of these methods are to the full extent described in monograph [1-4]. The light fullerenes molecules are very promising in the context of medical application, due to they demonstrate the following properties: antiHIV- protease activity, photodynamic DNA cleavage, free radical scavenger, antimicrobial action and use of fullerenes as diagnostic agents [1].

Light fullerenes (C_{60} and C_{70}) can be used in various areas of science and engineering, including materials science, mechanics, mechanical engineering, construction, electronics, optics, medicine, pharmacology, food and cosmetic industry [1,2]. However the application of light fullerenes is limited due to low solubility of fullerenes in water and aqueous solutions. For example, the solubility of C_{60} in water at 25 ^{o}C according to refs. 5-11 is equal to 1.3×10^{-11} $g \cdot l^{-1}$, and the solubility of C_{70} is equal to 1.1×10^{-13} $g \cdot l^{-1}$ [5-11]. Light fullerenes derivatives (fluoro, chloro, bromo, iodo, amino, carboxo, etc.) are also practically insoluble in water and aqueous solutions. However they have been extensively used in mechanical engineering (in water-soluble freezing and antifriction compositions), construction (as a soluble additives to cements and concretes), medicine and pharmacology (due to compatibility with water, physiological solutions, lymph, blood, digestive juices etc.), cosmetology (in the case of using of water and water + alcohol solutions) [1,2].

Fullerenols are the polyhydroxylated fullerenes, which contain different number of hydroxyl-groups.

During the hydroxylation of fullerenes mixtures of various polyhydroxylatied fullerenes can be obtained [12]. For example, hydroxylation of the C_{60} fullerene in the presence of quaternary ammonium bases results in $C_{60}(OH)_{26,5}$ – derivative, in the case of using HNO_3/H_2SO_4 – acids mixture the $C_{60}(OH)_{18\text{-}20}$ –derivative formation take place [13,14]. By the hydrolysis of the products of the RuO_4 and fullerenes reaction the diols 1,2-$C_{60}(OH)_2$, 1,2-$C_{70}(OH)_2$ and 5,6-$C_{70}(OH)_2$ can be synthesized. Due to the instability of such derivatives their application in biological experiments is impractible [15]. The $C_{60}(OH)_{24}$ fullerenol can be obtained by the reaction of alkaline hydrolysis of the polybromosubstituted $C_{60}Br_{24}$ fullerene [16, 23]. We conclude that the fullerenols obtaining by the methods presented in the literature leads to products of different structures and can be characterized by poor reproducibility and makes difficulties for experimental studies. Moreover, it has been shown that the fullerenol obtained by the reaction of the C_{60} fullerene with tertrabutylammonium hydroxide (TBAH) in toluene in the presense of oxygen and water solution of NaOH, is a stable anion-radical $Na_n^{+}[C_{60}O_xOH_y]_n^{-}$ (where: n = 2-3, x = 7-9, y = 12-15) [17]. Due to the low solubility of nonhydroxylated fullerenes, the rate of the direct reaction between C_{60} and hydrogen peroxide is rather slow, therefore it is more convenient to add the complementary hydroxy-groups to the hydroxylated fullerenol with 12 hydroxy-groups. Authors of ref. 18 carried out the reaction between the fullerenol with 12 hydroxy-groups and H_2O_2 (1) + H_2O (2) with ω_1=0.13 at 60 °C with shaking. Thus, the $C_{60}(OH)_{36}\cdot8H_2O$ and $C_{60}(OH)_{40}\cdot9H_2O$ fullerenols with different compositions and structures were synthesized over several days. Yang et al. [19] described the convenient and effective method of the fullerenol synthesis; polyethylene glycol 400 (PEG) as a catalyst was added to the reaction mixture containing fullerenes, aqueous solution of NaOH and an oxidizing agent. Sheng et al. [20,21] synthesized fullerenols by the direct reaction between the fullerene dredge and H_2O_2 + NaOH mixture. The eleven methods for the water-soluble fullerenols synthesis ($C_{60}(OH)_1$, $C_{60}(OH)_6$, $C_{60}(OH)_8$, $C_{60}(OH)_{x<12}$, $C_{60}(OH)_{12}$, $C_{60}(OH)_{x>15}$, $C_{60}O_{x<5}(OH)_{15}$, $C_{60}(OH)_{x<21}$, $C_{60}(OH)_{24}$ et al.) were described in the USA patent [21]. Authors of ref. 22, 23 synthesized the highly water soluble fullerenols by the direct oxidation reaction of the C_{60} fullerene under normal conditions. The fullerenols obtained retain the biological and chemical properties of the C_{60} fullerene; that allows using such derivatives in water-soluble oils, as additives to spirit production, in pharmacology and in medicine. Particularly, water-soluble films surfaces based on fullerenols can be used in micro- and optoelectronics, due to the following reasons:

- they can be extremely thin films (the thickness of such surfaces varies from 10 nm up to 1 μm);
- they have a high adhesion to metal, alloys and semi-conductors (A^3B^5, A^2B^6, A^4B^4 et al. types) surfaces; their transparency in visible and infrared spectrum regions is high; light-absorption in near-by ultra-violet spectral region is strong; in the far light-spectrum region such films are opaque; the refraction index is low; the chemical and thermostability is high. The fullerenol molecules have the potential bioapplicability as a free radical scavenger in biological systems, in oxidative stress induced by xenobiotics or radioactive irradiations; as an organo-protectors, neuroprotective (including Alzheimer's disease), non-specific analgetic, hepatoprotective and antiallergic agents [1,24].

1. METHODS OF INVESTIGATION OF PHASE EQUILIBRIA IN THE FULLERENE AND FULLERENOL – CONTAINING SYSTEMS

1.1. Methods of Determination of the Equilibrium Concentrations of Light Fullerenes and Fullerenol in Liquid Phase Using Gravimetric, Spectrophotometric Analysis and Liquid Chromatography Method

Experimental investigation of the individual light fullerenes and industrial fullerene mixture solubility in various biocompatible solvents was carried out by the isothermal saturation method [25-29]. First of all solutions of C_{60}, C_{70}, and fullerene mixture in corresponding solvent were prepared (in all cases the excess amount of solid phase was used). Then the solutions were stirred in a temperature-controlled shaker at particular fixed temperature for a period not shorter than 10 hours. The long duration of the experiment is connected with the long duration of the equilibrium state reaching in the concerned systems.

Calculations of the light fullerenes (C_{60}, C_{70}) concentrations in liquid solutions were realized on the basis of empirical equations obtained for fullerenes solutions in aromatic solvents by the spectrophotometric method (apparatus SF-4, Russia) [30-32]:

$$C(C_{60})=13{,}10(D_{335}-1.808D_{472}), \quad (1)$$

$$C(C_{70})=42.51(D_{472}-0.0081D_{335}), \quad (2),$$

where D_{335} and D_{472} represent the optical densities of solutions referred to the absorbance layer thickness equal to 1 cm; $C(C_{60})$, $C(C_{70})$ represent concentrations of the fullerenes in saturated solutions ($mg \cdot l^{-1}$). During the analysis admixtures of higher fullerenes we neglected [25-29].

In Figures 1-4 the absorbance spectra of light fullerenes in olive oil, beef fat, essential oil of carnation and linolenic acid in visible and UV-region are presented [25-29]. All the spectra shown are evidently typical for aromatic solvents (there are no solvatechromic effects and no displacements of absorption bands maxima) therefore the application of empirical equations (1, 2) is correct.

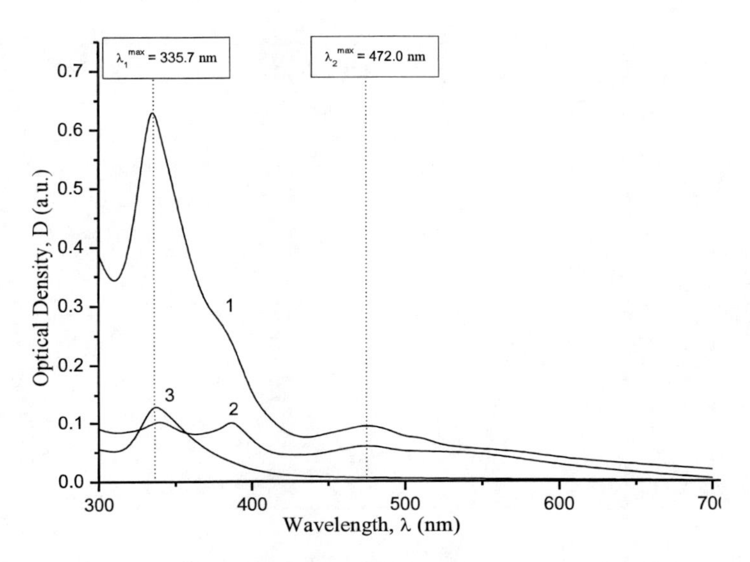

Figure 1. Optical spectra of light fullerenes in the olive oil: 1 – fullerene mixture (C60 – 65 %wt., C70 – 34 %wt., C76+C78+C84+C90… - 1 %wt.); 2 - C70 (99 %wt.); 3 - C60 (99.9 %wt.).

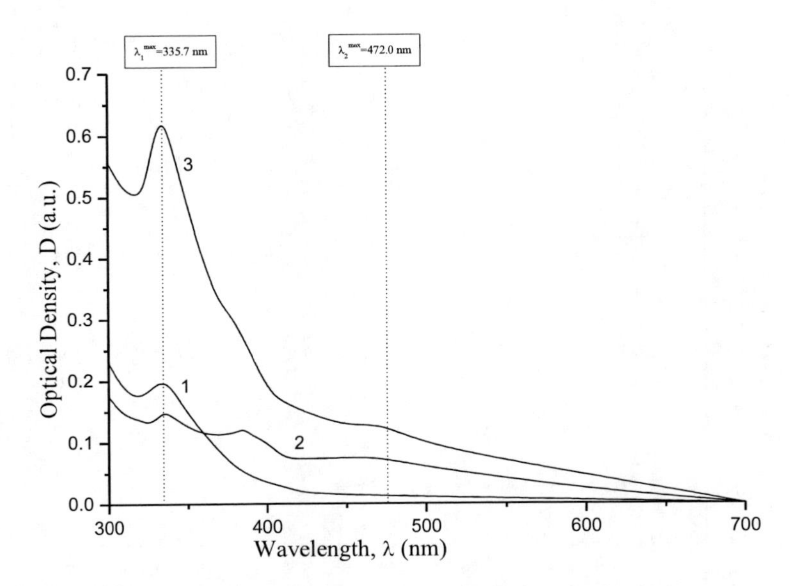

Figure 2. Optical spectra of light fullerenes in the beef fat: 1 – fullerene mixture (C_{60} – 65 %wt., C_{70} – 34 %wt., C_{76}+C_{78}+C_{84}+C_{90}… - 1 %wt.); 2 - C_{70} (99 %wt.); 3 - C_{60} (99.9 %wt.).

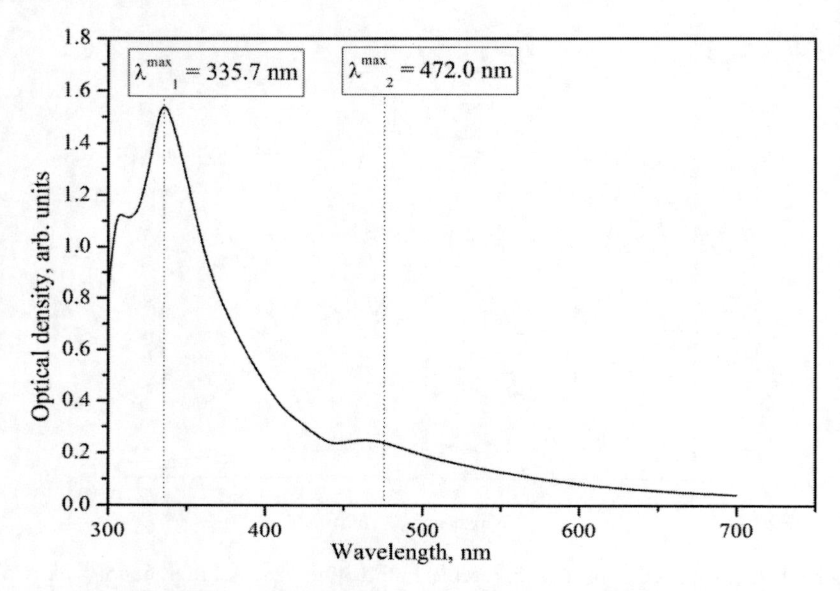

Figure 3. Optical spectra of light fullerene mixture C_{60} + C_{70} in essential oil of carnation.

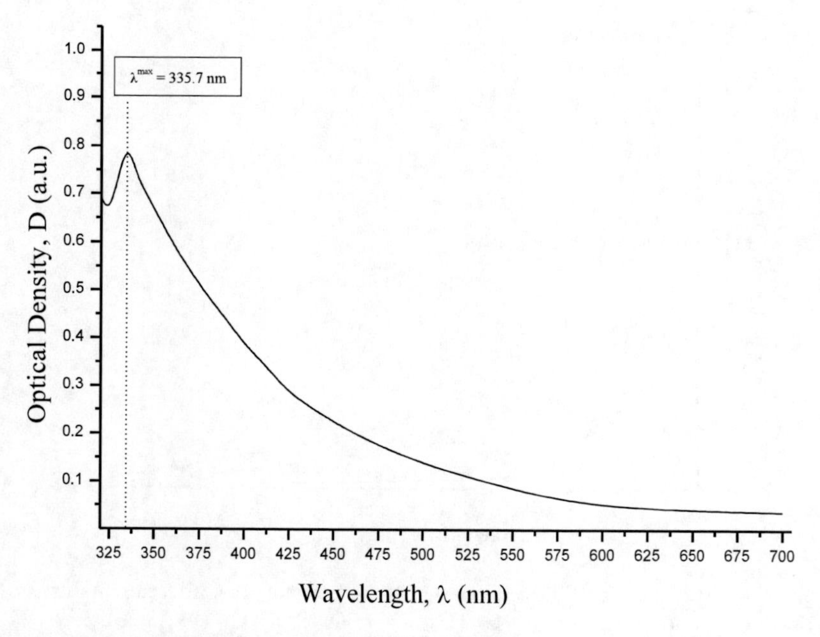

Figure 4. Optical spectra of the C_{60} fullerene in the linolenic acid (at 20^0C).

Authors of [26,27] analyzed the electronic spectra of the C_{60} fullerene solutions in olive oil, sunflower oil, linseed oil, biodiesel, soybean oil after heating at 150^{0}C (in air) for 15 minutes and revealed the formation of the new absorption band at 435 nm in the C_{60} spectrum; the couple hours stirring of solutions at 75^{0}C shows the same results [33,34]. The appeared absorption band at 435 nm was attributed to the 1,2-addition to the fullerene cage. The similar results were obtained by prolonged heating of the C_{60} fullerene solutions in nitrogen. On the contrary solutions of the C_{70} fullerene in vegetable oils showed the better stability. For example, only slight spectral changes were observed after 1 week ageing of the solutions of the C_{70} fullerene in methyl ester of brassica oilseed in air in one's turn heating of this solution under argon flow causes the profound changes in the visible part of the electronic spectra [33,34]. In the study [25,27] the investigation of the temperature dependences of solubility of the individual light fullerenes and industrial fullerene mixture were performed at lower temperatures (the chosen temperature range was 20-80^{0}C), than in mentioned above investigations; the duration of the fullerene's solutions was carried out in the closed volume under flow of argon. For this reason no changes in the electronic absorption spectra of the fullerene's solutions, except insignificant displacement of the absorption band at 335 nm wasn't observed.

The technique of studying of the light fullerenes solubility in cod-liver oil was absolutely identical to that presented above. Visible and near-UV absorption spectra obtained at saturation at 20°C are shown in Figure 5 [27]. Absorption spectra of fullerenes solutions in cod-liver oil shows the following:

- The basic general absorption peak of light fullerenes in UV range ($\lambda \approx 335.7$ nm in benzene, toluene, o-xylene) remains but is greatly displaced in UV area up to values of 308-311 nm;
- The second absorption peak of the C_{70} fullerene (but not C_{60}) and fullerene mixtures in UV range ($\lambda \approx 383$ nm) remains;
- The absorption peak of the C_{70} fullerene and fullerene mixtures disappears in visible area ($\lambda \approx 472$ nm).

All absorption peaks in UV areas in general disappear in absorption spectra obtained at saturation at 30°C in visible and near UV areas. At the same time, almost all the peaks corresponding to output of light fullerenes disappear, i.e., products of fullerenes interactions with cod-liver oil components do not go out from the chromatographic column at all.

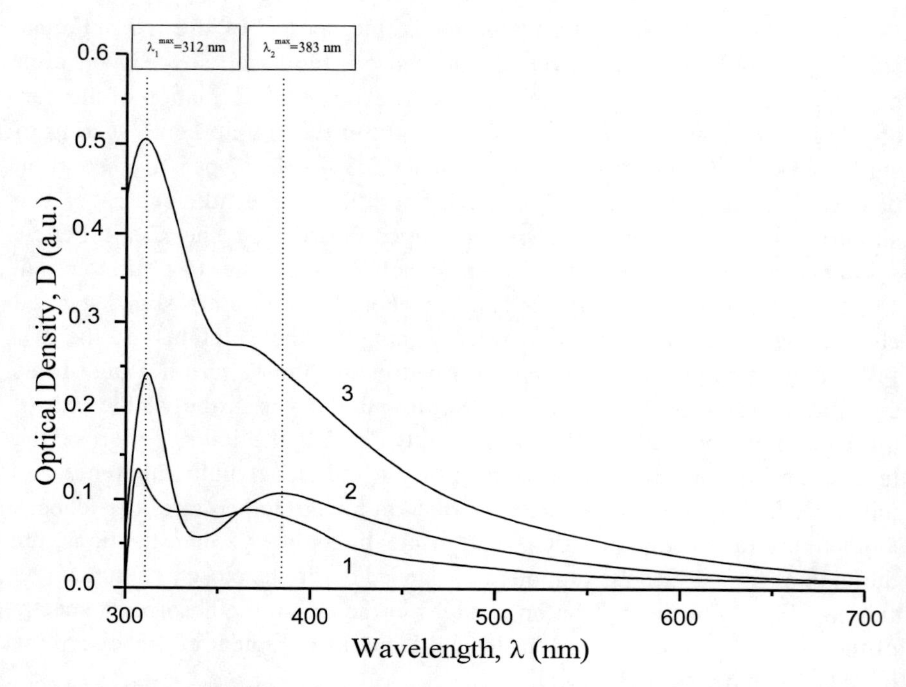

Figure 5. Optical spectra of light fullerenes in cod-liver oil: 3 – fullerene mixture (C_{60} – 65 %wt., C_{70} – 34 %wt., C_{76}+C_{78}+C_{84}+C_{90}… - 1 %wt.); 2-C_{70} (99 %wt.); 3- C_{60} (99.9 %wt.).

Thus we have to draw the conclusion that both the visible spectroscopy method and the liquid chromatography method (with spectrophotometric detecting at 254 nm) turned out to be unsuitable for the analysis of the fullerenes saturated solutions in cod-liver oil. It can probably be connected with irreversible formation of some constant composition compounds as a concequence of interaction of light fullerenes with some components of cod-liver oil (for example with components containing amino- and hydroxy-groups or two amino-groups simultaneously, etc.). In the case of cod-liver-oil, the light fullerenes solubility data could probably be obtained by the mass-spectrometry method.

Figure 3. shows visible and near UV absorption spectra of fullerene mixture C_{60} + C_{70} in the essential oil of carnation [28,29]. One can see from Figure 3. that the spectra are absolutely typical to light (C_{60} + C_{70}) fullerene solutions in aromatic solvents, which fact allows us to use the empirical formulas (1) and (2). But in some cases UV absorption spectra of individual light fullerenes or their mixture were destroyed and characteristic fullerenes peaks in spectra (≈ 335 and 472 nm) disappeared. Figure 6, as an example, shows visible and near UV absorption spectra of fullerene mixture $C_{60} + C_{70}$ in the essential oil of cedar [29]. In this case we only used the chromatographic method of the determination of light fullerenes concentrations. Thus, if the compound is soluble in a given component of the solvent, if the compound consists of two functional groups with heteroatoms (for example amino alcohol, diamine, amino acids, diols etc.) or if the compound contains the only functional group with heteroatom (for example amino or amide etc.) the distortion of the electronic spectra due to comparatively strong chemical interaction between light fullerenes and components of the solvent occur. Such kind of interaction influences on electron transition in the spectra of pristine fullerenes.

For examination of spectrophotometric data authors of ref. 25-29 used the liquid chromatography technique (liquid chromatograph «Lumachrom», St. Petersburg, Russia), the light-absorbance being detected at 254 nm. Results obtained by two independent methods are in a good agreement with each other. Accuracy of measuring of light fullerenes concentrations in saturated solutions was about ±(0.1-0.2) g/l depending on the current fullerene concentration.

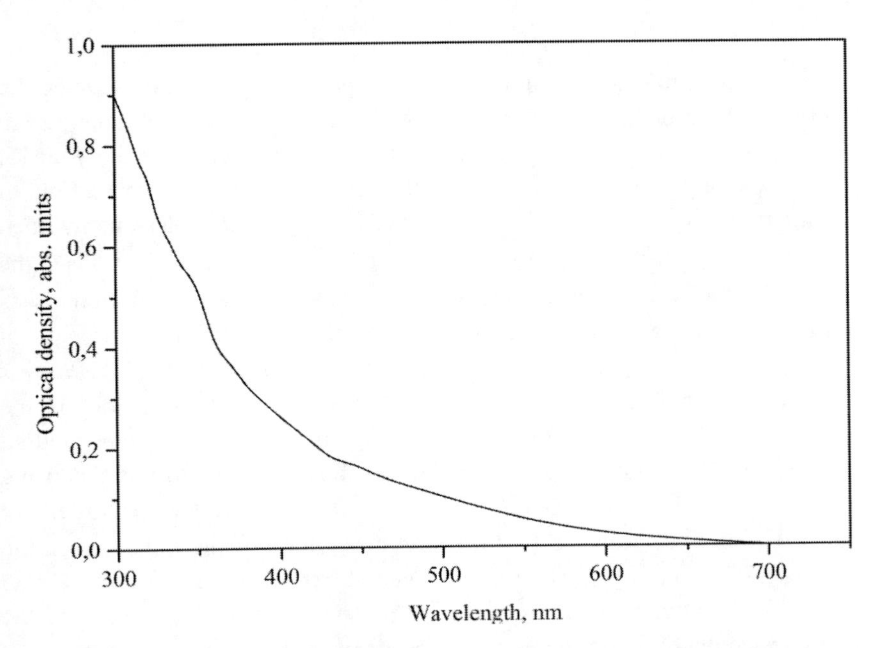

Figure 6. Optical spectra of light fullerene mixture $C_{60} + C_{70}$ in essential oil of cedar.

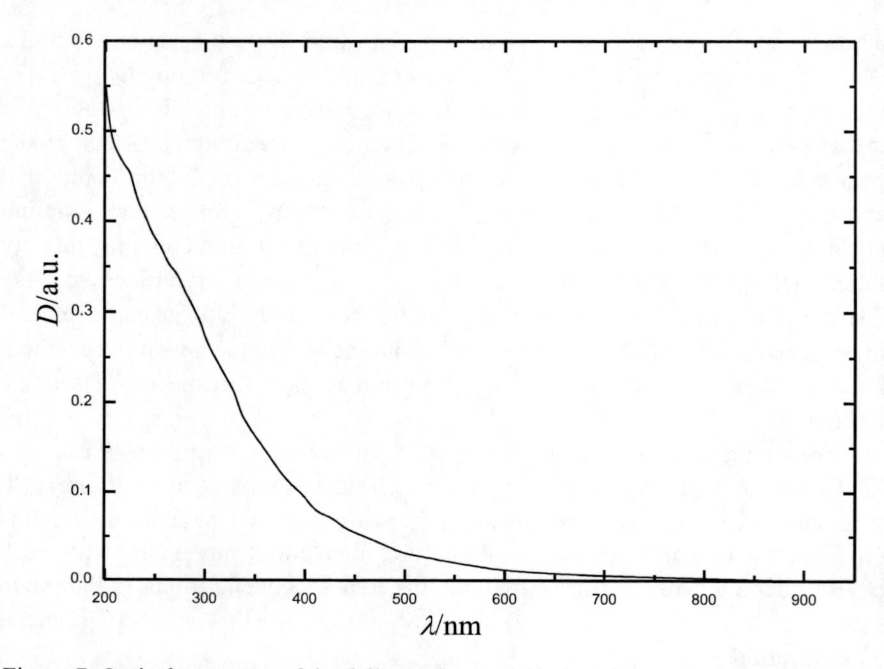

Figure 7. Optical spectrum of the fullerenol water solution. *D* is optical density; *λ*, wavelength.

By the isothermal saturation method in ampoules we have investigated the solubility of the fullerenol in the distilled water in the temperature range (20 to 80) °C [35-37]. The saturation conditions were the following: the time of saturation was equal to 120 min., the saturation of solutions was carried out in the temperature-controlled shaker (the saturation temperature was maintained with ± 0.05°C uncertainty), the shaking frequency was equal to 80 s^{-1}, the analysis of the fullerenol concentration in the saturated solutions was determined by the gravimetric method – the mass change in the process of evaporation of water from the solution at (50 ± 2) °C and 13.3 Pa was calculated. The concentration dependence of density determination of the fullerenol water solutions at 25 ^{0}C was carried out by the picnometer method [38] using quartz picnometers. The volume calibration was performed using water, the uncertainty of thermostatting was $\Delta t = \pm$ 0.1 K. The fullerenol water solutions were prepared by the direct dissolution of the fullerenol in the water. Then the heterogeneous mixture was stirred during 1 h at 25 ^{0}C and filtered. The mass concentration of the initial solution was determined by the gravimetric method. The other samples were prepared by the dilution of the basic solution. The fullerenol concentrations were calculated from the data on masses of fullerenol solutions and water.

Visible and near UV electron spectra of the aqueous solution of the fullerenol related to pure water are presented on Figure 7 [35-37]. The spectra were obtained on a SPECORD M-32 spectral photometer in quartz cuvettes of 1 cm width within the wavelength range of (200 to 900) nm. Figure 7 shows that the spectrum does not contain a visible absorption zone. In particular, no absorption peaks typical for the light fullerenes and their derivatives in aromatic and non- aromatic solvents were observed near 472 nm (for the fullerene C70), near 335 nm (for both fullerenes C60 and C70) and (320 to 330) nm (for bromine-fullerenes C60Brn (n = 6, 8, 24). As a whole, the UV-spectra of the fullerenol solutions turned out to be not informative enough, but can be used for the composition determination.

For example, the determination of the fullerenol composition is possible in aqueous environments if the wavelengths λ=(300 to 500) nm, where the absorption is not too high.

1.2. Methods of the Solid Solvates Composition Determination in the Systems Light Fullerenes, Fullerene Derivatives – Biocompatible Solvents, Using Thermogravimetric and Experimental Thermogravimetric Analysis

Determination of the possible composition of fullerene's solid solvates was carried out by the following way: a solid phase precipitated from a corresponding solvent was filtered on a Shott filter (porosity 10) washed out by ethanol and then dried during 30 minutes at 20-25°C. After that the solid phase was weighted, washed out by ethanol in Sokslet apparatus (78°C, 1atm), dried in vacuum (0.1 mm Hg., 200^0C) during 60 minutes, and then weighted again [25-29]. The composition of solid solvates of individual fullerenes (C_{60}, C_{70}) or solvated solid solutions was determined from the solid phase mass change. The mentioned above experimental method for determination of composition of solid phases was confirmed by the thermogravimetric analysis on Hungarian derivatograph Q-1500. The results of both methods are well agreed [25-27].

It makes sense to make some remarks concerning the possible compositions and dissociation temperatures of solid crystal solvates in systems light fullerenes - natural oils, animal fats [27]. The quantity of di- and monoglycerides, phospholipids, glycolipids, diolic lipids, free fatty acids, stearins and their esters, dye-stuffs, vitamins, polyphenols and their esters, and other substances which present as traces in the given oil is negligible, i.e., we consider the oil to consist of triglycerides entirely. Such assumption is quite reasonable as the content of triglycerides is actually high and in vegetable oils reaches up to 98 %wt. For hypothetic oil consisting of the only triglyceride (for example of the only triglyceride $TG \equiv R_1(CO)OCH_2 - CH(COOR_2) - CH_2O(CO)R_3$ having the mixed composition) the solid crystal solvate would have a strictly determined composition and a strictly determined dissociation temperature; the dissociation would occur according to the scheme: $C_{60} \cdot N(TG)_{(solid)} \rightarrow C_{60(solid)} + N(TG)_{(liquid)}$. However in the reality the situation described is impossible. Firstly, natural oils contain 5 and more various fatty acids (see Table1) which cannot be "packed" into one triglyceride in principle, and secondly, molar ratios of acids in such a triglyceride must be either 1:1:1 or 1:2 but it is never observed for the real natural oils (see Table 1).

Thus it should be accepted that oil is a multicomponent mixture of different triglycerides: $TG_1 + TG_2 + ... + TG_m$ (*m* is a number of various triglycerides in the particular oil). On the other hand it is clear that all such triglycerides must replace each other easily in the structure of solid crystal solvate for example in solid crystal solvate of the following composition:

$$KS \equiv C_{60} \cdot (TG_1)_{N1}(TG_2)_{N2}...(TG_{m-1})_{Nm-1}(TG_m)_{N-N1-N2-...-Nm-1}$$

owing to their structural and chemical similarity. Generally the number of triglycerides in the molecule of solid crystal solvate must be half-integer or small even number (N = 1/2 or 1 or 2 or 4 [9]. A similar picture should be observed for the solid crystal solvate of light fullerenes solid solutions:

$$KS \equiv (C_{60})_x(C_{70})_{1-x} \cdot (TG_1)_{N1}(TG_2)_{N2}...(TG_{m-1})_{Nm-1}(TG_m)_{N-N1-N2-...-Nm-1}$$

The composition analysis of the solid crystal solvates formed at relatively low temperatures ($T \leq 40^0$ C) (for example in olive oil) has shown that the sample weight loss of the solid crystal solvates after repeated washing by ethanol followed by drying in vacuum is $\Delta m \approx$ 18 ± 5 %wt. The average molecular weight of the mixed triglycerides which form a base of the olive oil is $\overline{M} \approx 885 \pm 15$ a. u. It was supposed in calculation that one average molecule of triglyceride in the olive oil contained 2.10 acid residua of oleic acid $CH_3(CH_2)_7CH{=}CH(CH_2)_7COOH$, 0.45 acid residua of palmitic acid $CH_3(CH_2)_{14}COOH$ and 0.45 acid residua of linoleic acid $CH_3(CH_2)_3(CH_2CH{=}CH)_2(CH_2)_7COOH$ (see Table 1).

And then simple procedure allows to calculate the average composition of the solid crystal solvate $KS \equiv C_{60}{\cdot}(0.17 \pm 0.05)TG$ where TG is a relative average molecule of triglyceride of the olive oil. Thus one acid residuum in triglyceride holds two molecules of C_{60}.

Similar results are obtained for the solid crystal solvates of C_{60} with other oils, for the solid crystal solvates of C_{70} and solid solutions based on solid crystal solvates of C_{60} and C_{70}.

Our general conclusion is the following: light fullerenes and solid solutions on their basis form the solid crystal solvates with natural vegetable oils which have rather low content of the solvent (in comparison with other solid crystal solvates); they are stable at the temperatures close to the room temperature [27].

The analysis of composition of the solid crystal solvates precipitated from solutions of the light fullerenes in the essential oils [28,29] and formed at relatively low temperatures ($T \leq 40^0$ C) has shown that loss of weight of the sample of the solid crystal solvates after repeated washing by ethanol followed by drying in vacuum is $\Delta m \approx 15 \pm 5$ mas. %.

Our general conclusion is as follows: light fullerenes and solid solutions based on them form the solid crystal solvents with essential oils which have rather low content of the solvent, viz. less than or equal to ¼ of «relative molecules of the essential oil» per one molecule of fullerene C_{60} (C_{70}) (in comparison with other solid crystal solvents) and are stable at the temperatures near the room temperature.

The mass change during elimination of oleic acid from the C_{60} solvate is equal to 2.0 % wt. and from C_{70} solvate is equal to 1.8 % wt. Taking into account the molecular weight of the oleic acid (282 a.u.) we can conclude that both of these values correspond to extremely low content of oleic acid in relevant solvates $C_{60}{\cdot}q[CH_3(CH_2)_7CH{=}CH(CH_2)_7COOH]$ and $C_{70}{\cdot}q[CH_3(CH_2)_7CH{=}CH(CH_2)_7COOH]$, $q = 0.05 \pm 0.02$. Thus we can consider both light fullerenes to be unsolvated in the solid phase. An analogous situation takes place in the case of the solid solvates of the light fullerenes with linoleic and linolenic acids [9].

Table 1. Thermal analysis of the fullerenol sample in the temperature range 20 - 1000 °C

№	Type of the effect	$t/\ ^0C$	Relative mass change $\Delta m / m \cdot 100$	Process
1.	TG	60-130	2 at 60°C, 10 at 75°C, 20 (19$^{theor.}$) at 100°C, 33 (32$^{theor.}$) at 120°C	$C_{60}(OH)_{22\div24}\cdot 30H_2O(solid) \rightarrow C_{60}(OH)_{22\div24}\cdot 12H_2O(solid) + 18H_2O(gas)$ $C_{60}(OH)_{22\div24}\cdot 12H_2O(solid) \rightarrow C_{60}(OH)_{22\div24}(solid) + 12H_2O(gas)$
2.	TG	820-950	40 at 830°C, 47 (47^{theor}) at 850°C, 58 at 900°C, 62 at 950°C	$C_{60}(OH)_{22\div24}(solid) \rightarrow C_{60}O_{11\div12}(solid) + (11\div12)H_2O(gas)$
3.	DTG	98	≈ 20 (19^{theor})	$C_{60}(OH)_{22\div24}\cdot 30H_2O(solid) \rightarrow C_{60}(OH)_{22\div24}\cdot 12H_2O(solid) + 18H_2O(gas)$
4.	DTG	119	≈ 33 (32^{theor})	$C_{60}(OH)_{22\div24}\cdot 12H_2O(solid) \rightarrow C_{60}(OH)_{22\div24}(solid) + 12H_2O(gas)$
5.	DTG	850	≈ 47 (47^{theor})	$C_{60}(OH)_{22\div24}(solid) \rightarrow C_{60}O_{11\div12}(solid) + (11\div12)H_2O(gas)$
6.	DTA	101	≈ 20 (19^{theor})	$C_{60}(OH)_{22\div24}\cdot 30H_2O(solid) \rightarrow C_{60}(OH)_{22\div24}\cdot 12H_2O(solid) + 18H_2O(gas)$
7.	DTA	120	≈ 33 (32^{theor})	$C_{60}(OH)_{22\div24}\cdot 12H_2O(solid) \rightarrow C_{60}(OH)_{22\div24}(solid) + 12H_2O(gas)$
8.	DTA	853	≈ 47 (47^{theor})	$C_{60}(OH)_{22\div24}(solid) \rightarrow C_{60}O_{11\div12}(solid) + (11\div12)H_2O(gas)$

For determination of the equilibrium fullerenol crystallohydrate $C_{60}(OH)_{22\div24}{}^{x}nH_2O$ composition the saturation of the fullerenol water solution in the temperature-controlled shaker at (25 ± 0.05) °C for 4 h was performed [35-37]. After that the solid phase was filtered and rapidly washed by methanol. The crystallohydrate obtained was weighed and dried in vacuum (13.3 Pa) at $t \approx (50 \pm 2)$ °C, and then weighed again. The relative mass change ($\Delta m_{fullerenol-\text{hydrate}} / m_{fullerenol-\text{hydrate}} \approx 0.30$) corresponded to the solvent content in the initial crystal solute. Thus the quantity of water molecules per one molecule of fullerenol is equal to 98 units. Thus the formula of the crystallohydrate is $C_{60}(OH)_{22\div24} \cdot 30H_2O$.

Additionally the thermogravic thermal analysis of the fullerenol crystallohydrate was performed (Figure 9, Table 1). The heating of the samples was carried in the open air with the heating rate equal to 5 K/ min. According to the thermal analysis data the water content in the equilibrium crystallohydrate $C_{60}(OH)_{22\div24} \cdot nH_2O$ is equal to *n* =*30* ± *2*. The fullerenol crystallohydrate dissociation can be expressed by the two-stage scheme:

$$C_{60}(OH)_{22\div24} \cdot 30H_2O(solid) \rightarrow C_{60}(OH)_{22\div24} \cdot 15H_2O(solid) + 15H_2O(gas)$$

$$t_1^{diss} \approx (100 \pm 2)^o C \quad (3.1),$$

$$C_{60}(OH)_{22\div24} \cdot 15H_2O(solid) \rightarrow C_{60}(OH)_{22\div24}(solid) + 15H_2O(gas)$$

$$t_2^{diss} \approx (119 \pm 2)^o C \quad (3.2).$$

With increasing of temperature the expansion of the fullerenol hydroxyls take place according to reaction:

$$C_{60}(OH)_{22\div24}(solid) \rightarrow C_{60}O_{11\div12}(solid) + (11 \text{ to } 12)\ H_2O\ (gas),$$

$$t_3^{diss} \approx (852 \pm 5)^o C \quad (3.3).$$

As an example the photos (magnification 21 and 300 times) of solid solvates of the fullerenole formed in binary system fullerenole + distilled water are presented on Figure 8. All of the photos were made on the polarizing microscope MIN-5 [37]

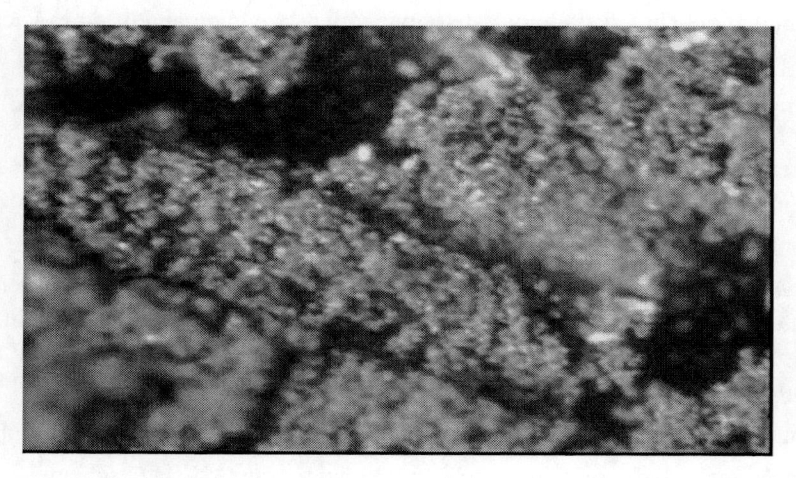

Figure 8.1.

Figure 8.2.

Figure 8. The photos (magnification 21 (1.1) and 300 (1.2) times) of the fullerenol crystals made on polarizing microscope MIN-5.

For the comparison of the thermal data obtained for the fullerenole with the corresponding data for the C_{60} fullerene the thermograms of the pristine C_{60} are presented additionally on Figure 9 [37].

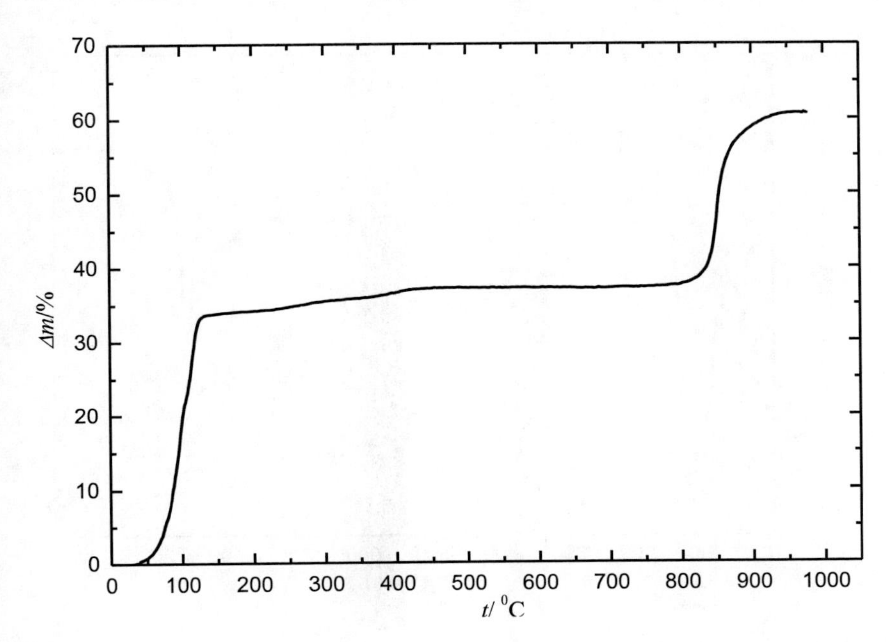

Figure 9.1.

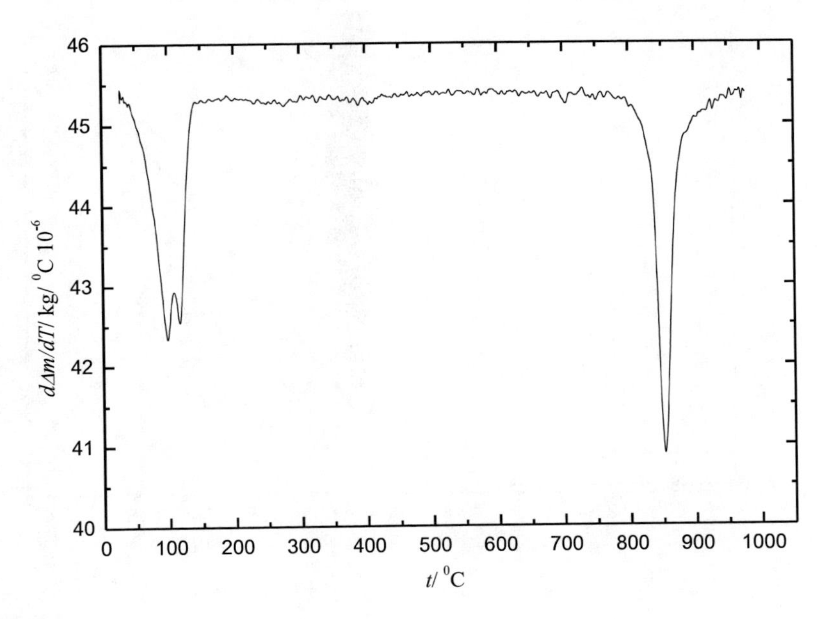

Figure 9.2.

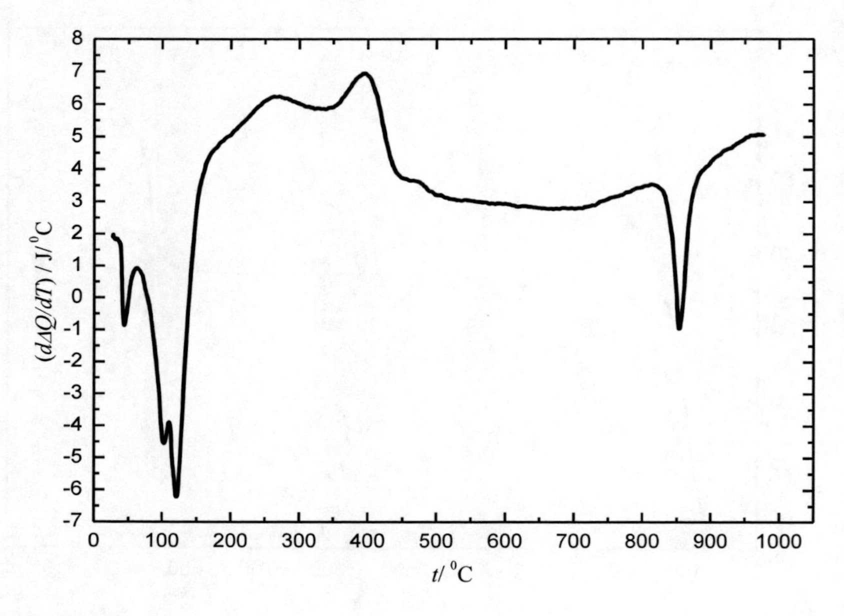

Figure 9.3.

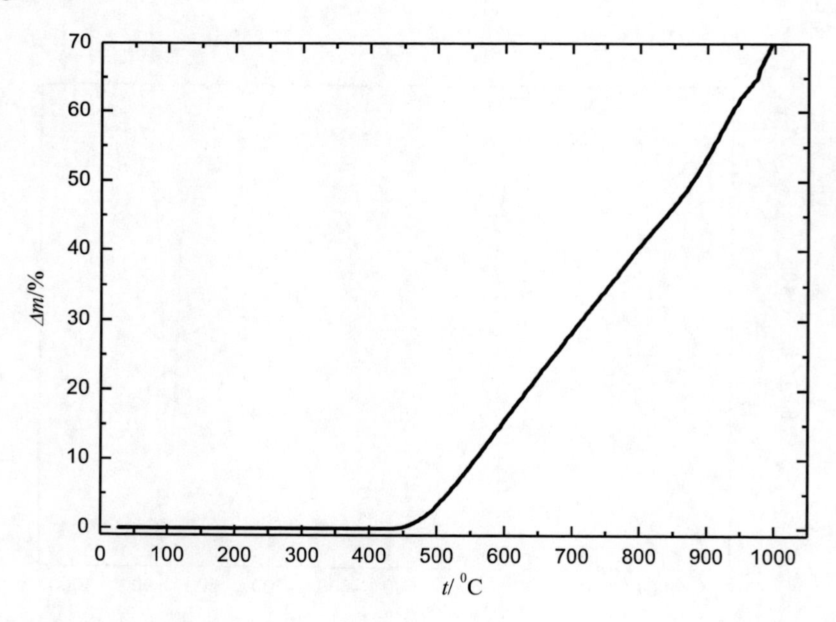

Figure 9.4.

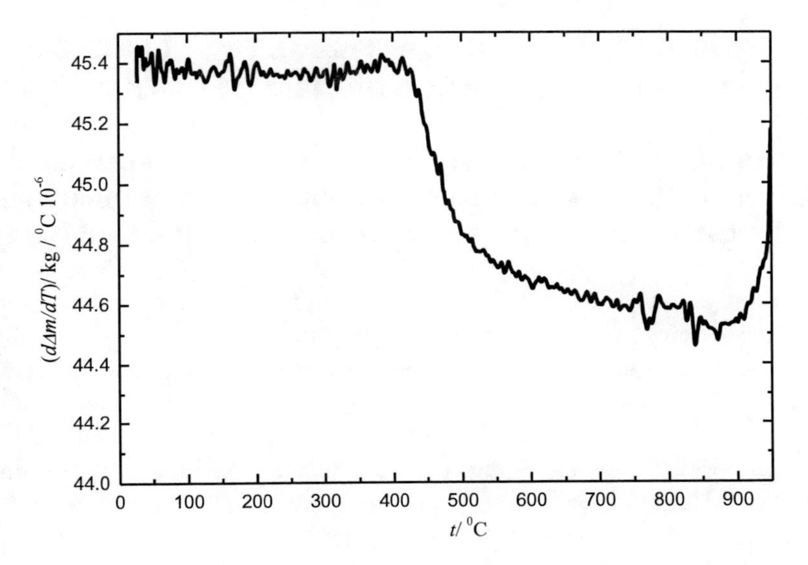

Figure 9.5.

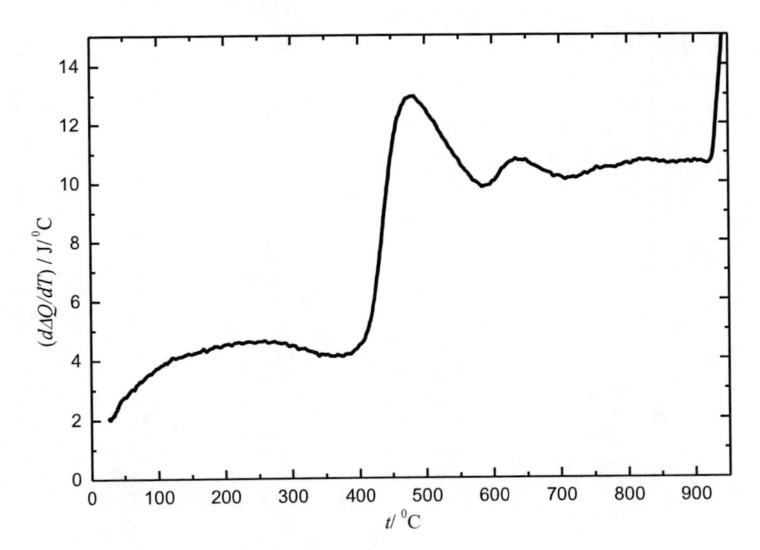

Figure 9.6.

Figure 9. Derivatograms of the fullerenole and individual C_{60} fullerene obtained in air at 5 ^{0}C/min heating rate; thermogravimetric curves (9.1 and 9.4), differential thermogravimetric curves (9.2 and 9.5), differential thermal analysis curves (9.3 and 9.6).

2. THE FULLERENOLE AND LIGHT FULLERENES SOLUBILITY IN BIOCOMPATIBLE SOLVENTS

2.1. Solubility of Light Fullerenes and Industrial Fullerene Mixtures in Polyunsaturated Acids, Natural Oils, Animal Fats and Essential Oils. Ways of Application of the Obtained Data

Investigation of the light fullerenes solubility in unsaturated fatty acids is actual because of the compositions including light fullerenes and unsaturated fatty acids can be used in the elaboration of tribological additives to fuels and technical oils. It is significant that the compositions including unsaturated fatty acids mainly based on oleic acid (due to its low price) are successfully used for these purposes [9]. On the Figure 1 the temperature dependences of solubility of the individual light fullerenes C_{60} (Figure 10.1) and C_{70} (Figure 10.2.) in oleic, linoleic and linolenic acids are presented.

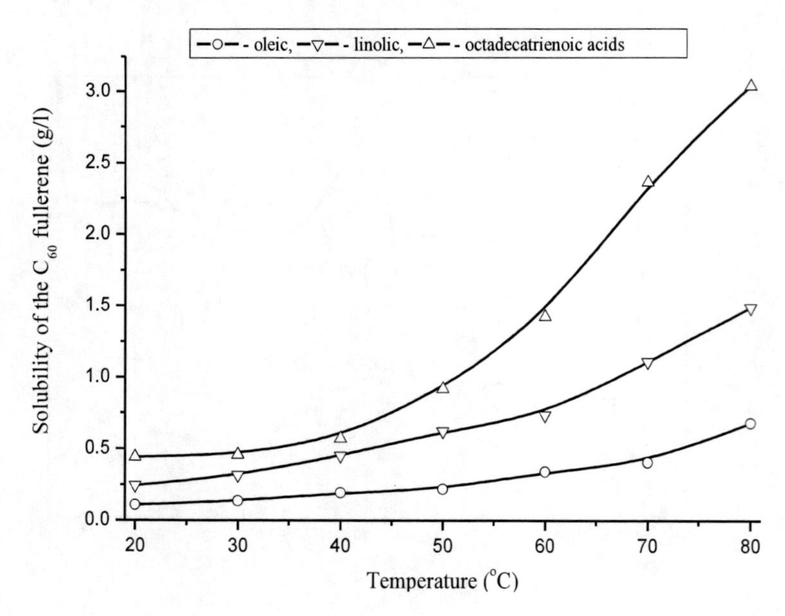

Figure 10.1.

The solubility polytherms of the components from the industrial fullerene mixture (namely C_{60} and C_{70}) and the total solubility of the industrial fullerene mixture components are presented on the Figures 11.1, 11.2 and 11.3 correspondingly [9].

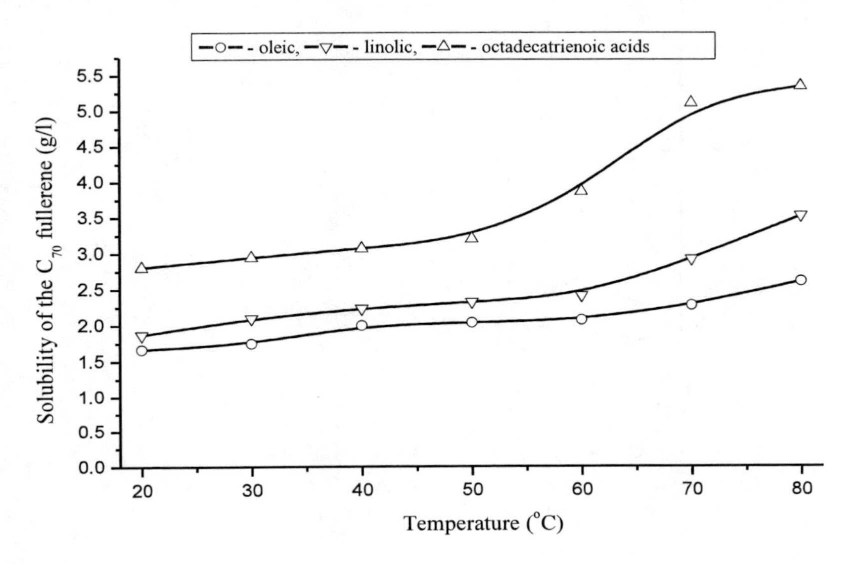

Figure 10.2.

Figure 10. Solubility of individual light fullerenes C_{60} (10.1.) and C_{70} (10.2.) in oleic, linolic and octadecatrienoic acids in the temperature range 20-80°C.

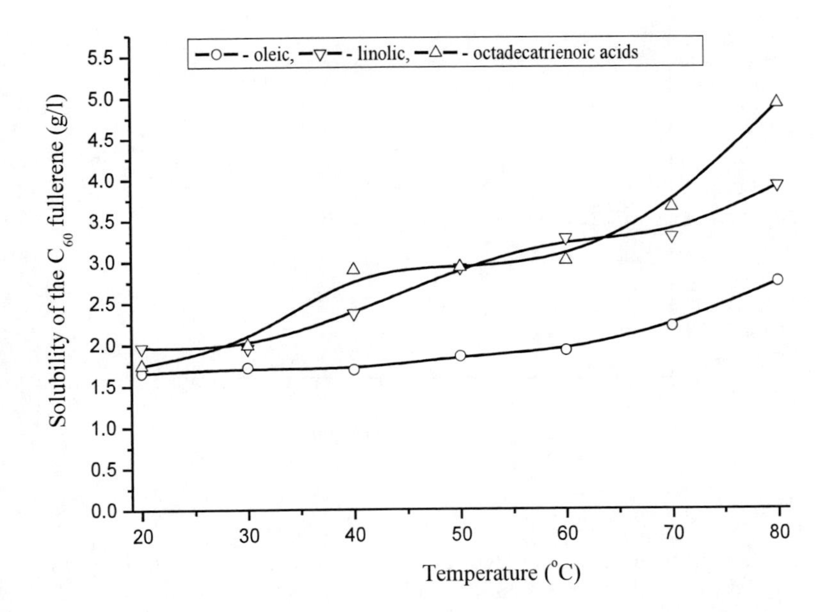

Figure 11.1.

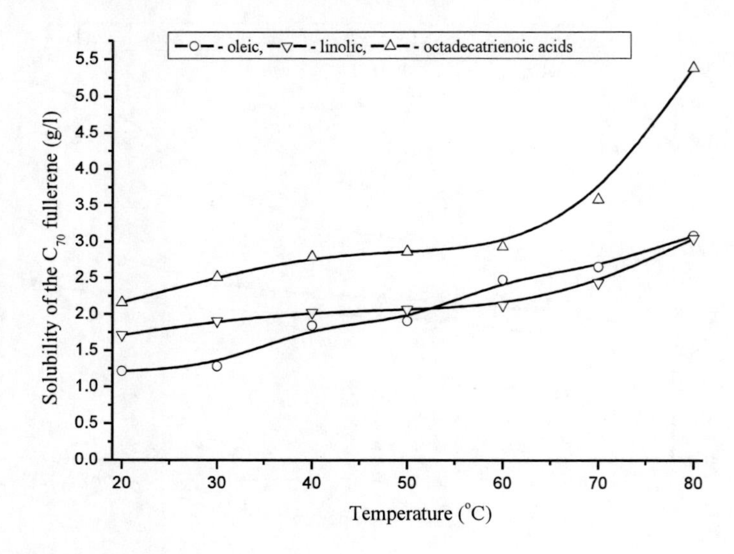

Figure 11.2.

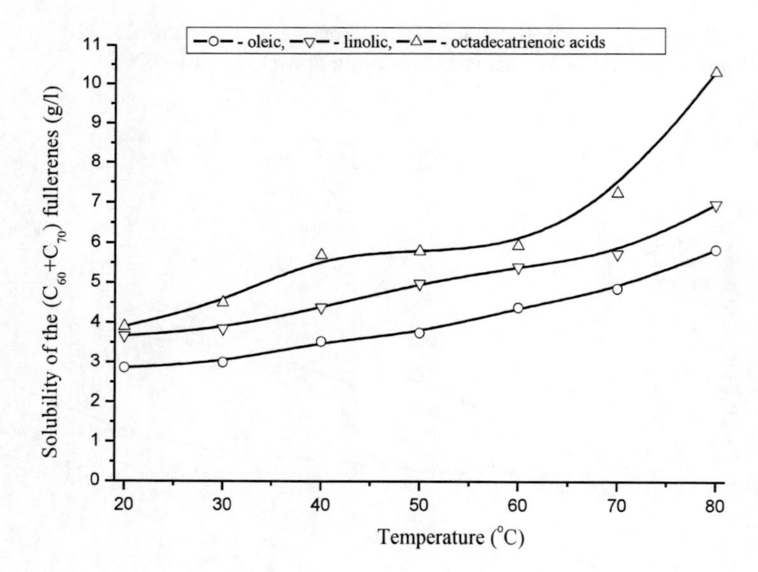

Figure 11.3.

Figure 11. Solubility of the C_{60} fullerene (11.1.), C_{70} fullerene (11.2.) from industrial fullerene mixture and the total solubility (C_{60}+C_{70}) of the fullerene mixture (11.3.) in oleic, linolic and octadecatrienoic acids in the temperature range 20-80°C.

Figures 10.1,10. 2 reveal the following [9]:

- solubility in all studied unsaturated fatty carboxylic acids varies from hundredth mg/l up to 10 g/l. It is worth to note that these values are 1-2 orders higher than in the case of earlier studied saturated carboxylic acids;
- solubility of both light fullerenes in all studied unsaturated carboxylic acids increases with increasing temperature (in the case of individual light fullerenes (C_{60}, C_{70}) and components of industrial fullerene mixture): $dS_{acid}(C_{60})/dT>0$ and $dS_{acid}(C_{70})/dT>0$;
- in all cases the individual solubility of the C_{70} fullerene is higher than the solubility value of the "pseudospheroid" C_{60} fullerene. This fact is quite typical for the solvents where solubility of light fullerenes is high (1-5, 8): $S_{acid}(C_{60})<S_{acid}(C_{70})$. The phenomenon of higher solubility of C_{70} is determined by higher polarizabillity of this fullerene ($8*10^{-23}cm^3$) in comparison with C_{60} ($6,4*10^{-23}cm^3$);
- the solubility values of the light fullerenes from industrial fullerene mixture (65% C_{60}, 39% C_{70}, 1% C_{76-90}) are similar: $S_{acid}(C_{60})\approx$ $S_{acid}(C_{70})$;
- solubility of the C_{60} from the industrial fullerene mixture is higher than the solubility of individual C_{60}; the "salting-in" effect of the C_{60} fullerene is observed in the branch of crystallization of the solid solutions enriched with C_{60};
- solubility of the C_{70} fullerene from the fullerene mixture almost coincides with the solubility of individual C_{70}. We have observed such kind of effects earlier in the ternary systems with the aromatic solvent o-xylene;
- with increasing of the unsaturation degree of a carboxylic acid the solubility of both light fullerenes increases: solubility of fullerenes in trienoic acid is higher than in dienic acid and solubility values in dienic acid are higher than in "monoenic" acid:

$$S_{linolenic}(C_{60})>S_{linoleic}(C_{60})>S_{oleic}(C_{60})\,,\; S_{linolenic}(C_{70})>$$
$$S_{linoleic}(C_{70})>S_{oleic}(C_{70})\,.$$

The increasing of the solubility values of fullerenes with increasing of unsaturation value of investigated fatty acids is regular. The fundamental characteristics of fullerene molecules are the first electron affinity (EA) and the first ionization energy (IE). For C_{60} EA=(4.273±0.002) · 10^{-19} J, IE=(12.13±0.02) 10^{-19} J [2].

Notable that the value of electron affinity of C_{60} is higher than that of aromatic solvents [39]. This illustrates the strong electron-acceptor properties of fullerenes. Analysis of the available literature data reveals that fullerenes are known to have strong electron transfer interactions with electron donors such as aromatic solvents, aliphatic amines [39]. Thus, we deal with the Lewis acid character of fullerenes. For confirming of the latest fact we can consider the changes of the solubility values in different classes of organic solvents depending on number of unsaturated bonds: for the C_{60} fullerene – n-hexane (0.037 g/l), cyclohexane (0.36 g/l), cyclohexene (1.21 g/l), benzene (1.5 g/l) [9]; for the C_{70} fullerene - hexane (0.013 g/l), benzene (1.3 g/l) [9]. We can conclude that increasing of the number of double bonds in the structure of the solvent cause the more strong $\pi-\pi$ interaction between electron deficient fullerene molecules compound and the molecules of the unsaturated compounds rich in π - electrons. Addition of the electrono-donor groups with plus inductive effect (alkyl-radicals) or plus mesomeric effect (halogens, amino-groups, hydroxyl-group etc.) to the structure of solvents also increase the solubility value of fullerene. The process of a fullerene dissolution involves several stages. These stages are accompanied with the following energy effects: endothermic destruction of the fullerite lattice, breaking of solvent-solvent bonds, as well as compensative effects of exothermic solvation process. When the potential energy of solvent and solute molecules in the electrostatic field of solution are similar (Semenchenko's rule) the solubility of a solute reaches its maximum because of strong solute-solvent interactions [40]. Ruoff et al. demonstrated that there are no molecular parameters which can universally predict the solubility of fullerenes however they determined a set of solvent parameters which can predict well fullerene solubility [41]. These solvent characteristics are the following: high value of refraction index; value of dielectric permittivity equal to 4; high molecular volume of the solvent; value of Hildebrand's solubility parameter [44-46] equal to 20 $\left[\frac{J}{cm^3}\right]^{\frac{1}{2}}$; nucleophilic properties of a solvent [41,42]. The literature data on dielectric permittivities, refraction indexes and molecular volumes of oleic, linoleic and linolenic acids in comparison with the isothermal solubility values (at 20^0C) of the individual light fullerenes are presented in Table 2 (the values of the Hildebrand's parameters [43] are absent in literature due to the reason of unavailability of the evaporation enthalpies of the fatty acids; we used the values obtained in [33] based on using Van Krevelen method).

Table 2. Molecular characteristics of the oleic, linoleic and linolenic acids [33,47]

Solvent	δ, $\left[\frac{J}{cm^3}\right]^{\frac{1}{2}}$	ε (20^0C)	n (20^0C)	S (C_{60}), g/l (20^0C)	S (C_{70}), g/l (20^0C)
oleic acid	18.8	2.46	1.45823	0.109	1.667
linoleic acid	18.7	2.6	1.4699	0.244	1.867
linolenic acid	18.7	2.76	1.4808	0.442	2.804

Analysis of the data presented in Table 2 shows the increasing of the solubility values of fullerenes with increasing of the values of the molecular characteristics of the solvents (dielectric permittivities, refraction) in the series oleic acid – linolenic acid. This fact is well agreed with the mentioned above Fuoff's empirical rules guarantee the high solubility values of fullerenes.

The authors of [48-51] studied the isothermal solubility of the individual light fullerenes (C_{60} and C_{70}) in various oils and fats (linseed, sunflower, soybean, olive, castor, peanut, sunseed oils and molten cow butter); the theoretical research on prediction of solubility in such multicomponent systems based on Hansen solubility model was presented [33]. The correlation between the solubility value of the C_{60} fullerene and unsaturation level of the vegetable oil was found. This correlation reveals that esters with saturated fatty acids are better solvents for fullerenes than esters with increasing levels of double bonds [33].

Fats (oils) are known to be the substances of animal, vegetable or microbial origin consisting of about 98 %wt. triglycerides (absolute ethers of glycerin and fatty acids). They also consist of di- and monoglycerides (1–3 %wt.), phospholipids, glycolipids, diol lipids (0.5-3 %wt.), and traces of uncombined fatty acids, stearins and their ethers, as well as dye-stuffs, vitamins A, D, E, K, polyphenols and their ethers [52]. In general chemical properties of fats depend on the carbon chain length, degree of fatty acids unsaturation, and their arrangement in triglycerides. Commonly fats (oils) consist of straight-chain fatty acids (saturated, unsaturated and polyunsaturated) with even number of carbon atoms (usually $4 \div 26$).

It should be noted that the most part of unsaturated acids of natural fats (oils) is located in the β-position of triglycerides while the saturated acids are located mainly in the α-position of triglycerides [52]. The fatty acid composition of natural oils which were used in the present study is shown in the Table 3 whereas the composition of fats is presented in the Table 4.

Investigation of the light fullerenes solubility in liquid natural vegetable oils and animal fats is extremely important due to the following reasons:

- fullerenes are rather well soluble in this natural solvents in the mentioned temperature range. The solubility varies from tenth to units gram of fullerenes per liter of solvent;
- fullerenes solutions in natural solvents represent stable and absolutely transparent true solutions;
- these solutions are absolutely nontoxic and biocompatible with respect to animals and humans only in the case they are prepared by extraction of fullerene mixture from fullerene soot by natural oils and adipose, i.e., in the said case they do not contain any impurities and unhealthy components. On the contrary if one uses standard individual fullerenes obtained from solutions in aromatic solvents (toluene, o-xylene, dichlorobenzene) such fullerenes contain residuals of these solvents. These residuals are retained even after drying a fullerene under vacuum at 0.01 mm Hg and 200 – 250°C, their content varying from hundredth to thousandth mass percent. Alternative way for the complete purification of a fullerene from solvents admixtures is a high-temperature sublimation of a fullerene at low pressure ($<10^{-5}$ mm Hg.) but the latter process is very expensive and laborious;
- fullerenes solutions in oils and fats possess pronounced antibacterial and antioxidative properties, they can absorb free-radicals and ion-radicals from condensed phases in which they are present as well as photons in ultra-violet area of the spectrum.

Table 3. Average fatty acid composition of natural oils

Sort of natural oil	Content of fatty acids (%wt.)			
	lauric	myristic	palmitic	palmitooleic
corn oil	<0.3	<0.3	9.0-14	<0.5
grape-kernel oil	-	<0.2	5.6-7.6	<0.3
olive oil	-	-	7.0-20.0	0.3-3.5
peach-kernel oil	-	<0.1	<3.0	-
sunflower oil	-	<0.2	5.0-7.6	<0.3
cedar oil	-	-	3.0-3.9	-
walnut oil	-	-	8.0	-
linseed oil	-	-	7.0	-
Sort of natural oil	Content of fatty acids (%wt.)			
	stearic	oleinic	linoleic	octadecatrienoic
corn oil	0.5-4.0	24-42	34-62	-
grape-kernel oil	2.7-6.5	14.6-39.4	18.3-74.0	-
Sort of natural oil	Content of fatty acids (%wt.)			
	lauric	myristic	palmitic	palmitooleic
olive oil	1.5-4.3	56.0-86.0	3.3-20.0	0.4-1.5
peach-kernel oil	10.0-15.0	73.3-85.0	5.0-16.4	-
sunflower oil	2.7-5.5	14.0-39.4	48.3-77.0	<0.3
cedar oil	3.4-4.1	22.1-26.0	36.99-69.0	18.0-24.3
walnut oil	2.0	20	56	14
linseed oil	3.0	25	12	50
corn oil	<1.0	<0.5	-	<0.5
grape-kernel oil	0.2-0.4	<0.2	-	0.5-1.3
olive oil	0.2-1.6	0.2-0.5	-	-
peach-kernel oil	-	-	-	-
sunflower oil	<0.5	<0.3	-	0.3-1.5
cedar oil	<0.3	0.8-1.3	<0.4	-

Table 3. (Continued)

Sort of natural oil	Content of fatty acids (%wt.)			
	arachidonic	gondoinic	eicosapentaenoic	behenic
walnut oil	-	-	-	-
linseed oil	-	-	-	-
Sort of natural oil	Content of fatty acids (%wt.)			
	erucic	lignoceric		
corn oil	<0.5	-		
grape-kernel oil	0.5-1.3	<0.2		
olive oil	-	-		
peach-kernel oil	-	-		
sunflower oil	0.3-1.5	-		
cedar oil	-	-		
walnut oil	-	-		
linseed oil	-	-		

"-" denotes a negligible value or that the composition isn't defined.

Table 4. Average fatty acid composition of animal fats

Sort of animal fat	Content of fatty acids (%wt.)			
	myristic	palmitic	stearic	9-hexadeceneonic
beef fat	3.0-3.3	24-29	21-24	2.1-2.7
lamb fat	2.2-3.0	23-30	20-31	12-13
pork fat	0.8-0.9	27-30	13-18	1.7-1.9
bird fat	0.8-1.7	20-26	4.0-9.0	3.0-9.0
milk fat	8.0-17	24-29	9.0-13	4.0
codliver oil	4.5-11	10-29	0.7-4.0	5.0-20
Sort of animal fat	Content of fatty acids (%wt.)			
	oleinic	linoleic	9-eicosenic	11,14-eicosadienic
beef fat	41-42	2.0-5.0	-	-
lamb fat	35-41	3.0-4.0	-	-
pork fat	37-44	8.0-9.0	-	-
bird fat	33-46	10-22	-	-
milk fat	19-34	2.0	-	-
codliver oil	7.0-26	0.5-3.0	0.3-20	0.0-2.0
Sort of animal fat	Content of fatty acids (%wt.)			
	arachidonic acid	docosahexaenic		
codliver oil	0.3-1.0	0.4-12		

"-" denotes a negligible value or that the composition is not defined.

Table 5. Solubility of light fullerenes in natural oils [26,27]

Solubility of C_{60}, g/l					
Solvent	0^0C	20^0C	40^0C	60^0C	80^0
«Stavropolje» (unrefined sunflower oil)	0.294	0.451	0.727	0.708	0.638
«Zlato» (refined sunflower oil)	0.299	0.377	0.456	0.570	0.697
«Milora» (corn oil)	0.196	0.609	0.846	0.945	1.319
walnut oil	0.269	0.485	0.694	0.718	1.017
olive oil	0.264	0.470	0.751	0.858	1.206
linseed-oil	0.229	0.513	0.639	0.717	1.033
pignolia oil	0.323	0.485	0.818	1.053	1.099
grape-seed oil	0.327	0.371	0.663	0.748	0.897
apricot - kernel oil	0.372	0.509	0.702	0.799	0.958
«Stavropolje» (unrefined sunflower oil)	0.504	1.827	2.647	2.872	3.213
«Zlato» (refined sunflower oil)	0.303	1.289	1.668	2.996	3.261
«Milora» (corn oil)	0.506	1.957	2.963	3.404	4.286
walnut oil	0.315	1.390	1.829	2.650	2.837
olive oil	0.402	0.756	1.286	2.735	3.278
linseed-oil	0.505	1.198	1.639	2.118	2.773
pignolia oil	0.378	0.757	1.512	2.020	2.710
grape-seed oil	0.632	0.947	1.578	2.207	2.838
apricot - kernel oil	0.378	1.009	2.208	2.772	3.090
Solubility of the fullerene mixture (65 %wt. C_{60}, 34 %wt. C_{70}, 1 %wt. C_{76-90}), g/l; upper value corresponds to the solubility of C_{60} from fullerene mixture; lower value corresponds to the solubility of C_{70} from fullerene mixture					
«Stavropolje» (unrefined sunflower oil)	1.831	2.234	2.068	1.839	1.612
	0.898	1.045	1.361	1.843	2.320
«Zlato» (refined sunflower oil)	1.828	2.284	2.510	2.663	2.864
	1.279	1.456	1.731	1.825	2.199
«Milora» (corn oil)	1.949	2.294	2.746	2.619	2.735
	1.209	1.201	1.346	1.664	1.976
walnut oil	1.378	2.127	2.156	2.264	2.340
	0.752	0.890	1.362	1.517	1.546
«Stavropolje» (unrefined sunflower oil)	1.831	2.234	2.068	1.839	1.612
	0.898	1.045	1.361	1.843	2.320
«Zlato» (refined sunflower oil)	1.828	2.284	2.510	2.663	2.864
	1.279	1.456	1.731	1.825	2.199
«Milora» (corn oil)	1.949	2.294	2.746	2.619	2.735
	1.209	1.201	1.346	1.664	1.976

Table 5. (Continued)

Solubility of C_{70}, g/l					
walnut oil	1.378	2.127	2.156	2.264	2.340
	0.752	0.890	1.362	1.517	1.546
olive oil	1.239	2.900	2.224	2.135	2.150
	0.913	1.123	1.518	1.747	1.999
linseed-oil	1.733	2.097	2.067	1.987	1.923
	0.900	1.117	1.207	1.524	1.778
pignolia oil	1.711	2.026	2.479	2.484	2.607
	1.370	1.681[b]	1.826	2.079	2.294
grape-seed oil	1.693	1.801	2.092	2.263	2.224
	1.059	1.214	1.301	1.517	1.675
apricot - kernel oil	1.437	1.791	1.905	2.086	2.097
	0.908	1.056	1.463	1.487	1.553

The polythermal solubility curves of fullerenes C_{60} and C_{70} in vegetable oils within the temperature range from 0 to 80°C are shown in Figures 12, 13 and Table 5 [26,27].

Figures 12, 13 show the following:

- The solubility increases with the temperature increase from 0°C up to 80°C for both C_{60} and C_{70} fullerenes in all natural oils (by 2-3 times for the C_{60} fullerene and by 5-10 times for the C_{70} fullerene)
- The C_{70} fullerene solubility in considered natural oils is always higher than the C_{60} fullerene solubility in the same oils (as a rule by 1.5-2.0 times at 0°C and by 3-5 times at 80°C)
- The highest solubility for both fullerenes at high temperatures is observed for corn oil.

The polythermal solubility curves for the fullerenes sum (C_{60}+C_{70}) from the fullerene mixture (60 %wt. of C_{60}, 39 %wt. of C_{70}, 1 %wt. of C_{76-90}) in natural oils are shown in Figure 14 and Table 5 (during analysis, negligible admixtures of the higher C_{76-90} fullerenes in given fullerene mixtures were ignored) [26,27]. The solubility of the fullerenes sum (C_{60}+C_{70}) also monotonously increases with the temperature increase from 0 up to 80°C from 2-3 g l^{-1} up to 3-5 g l^{-1} depending on the type of natural oil as shown in Figure 14.

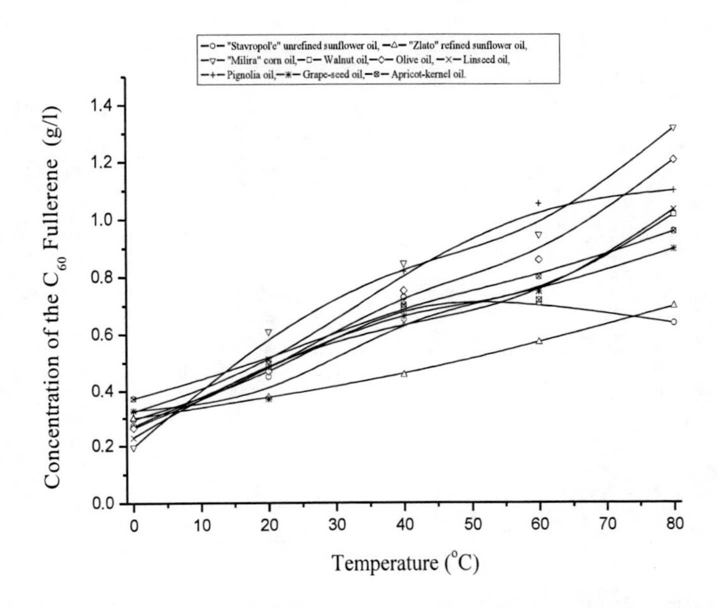

Figure 12. Polythermal solubility of C_{60} in natural oils in the temperature range 0 – 80^{0}C.

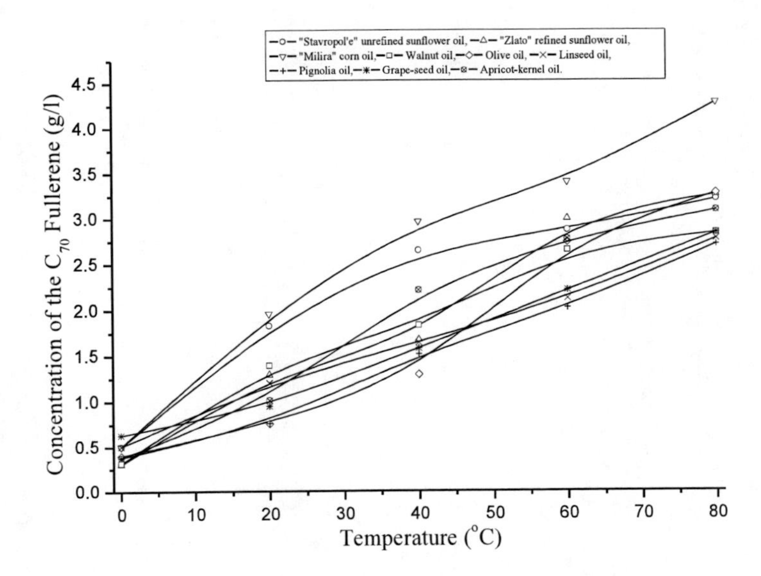

Figure 13. Polythermal solubility of C_{70} in natural oils in the temperature range 0 – 80^{0}C.

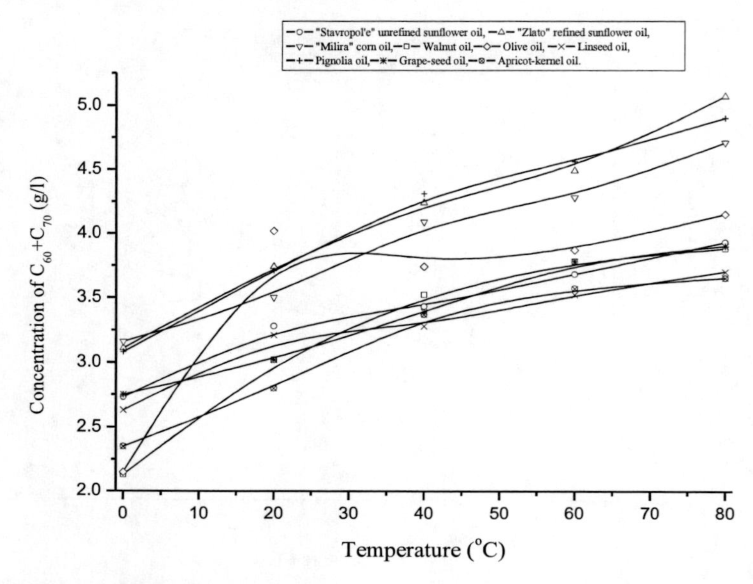

Figure 14. Polythermal solubility of the industrial fullerene mixture (C_{60} – 65 %wt., C_{70} – 34 %wt., C_{76}+C_{78}+C_{84}+C_{90}… - 1 %wt.) in natural oils.

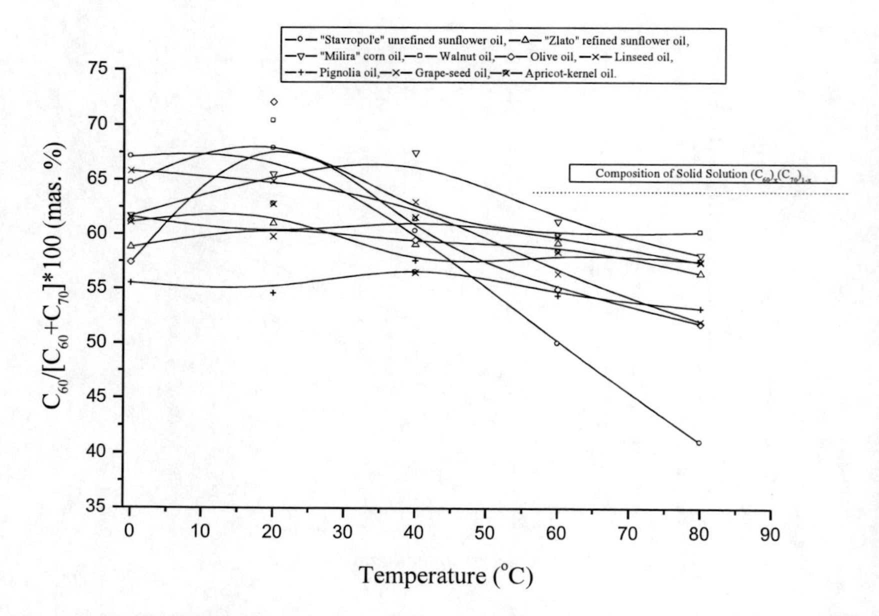

Figure 15. Fullerene content X(C_{60}) in saturated solution of fullerene mixture in natural oils against temperature. The dotted line corresponds to the initial solid solution.

Table 6. Solubility of light fullerenes in animal fats

Solubility of C_{60}, g/l				
Solvent	50^0C	60^0C	70^0C	80^0C
pork fat	0.052	0.145	0.231	0.257
bird fat	0.024	0.059	0.093	0.122
beef fat	0.041	0.131	0.153	0.168
margarine «Rumjashka»	-	0.019	0.028	0.029
lamb fat	0.041	0.212	0.385	0.541
desi	0.021	0.046	0.063	0.087
Solubility of C_{70}, g/l				
Solvent	50^0C	60^0C	70^0C	80^0C
pork fat	0.143	0.291	0.529	0.582
bird fat	0.148	0.344	0.661	0.688
beef fat	0.054	0.096	0.134	0.160
margarine «Rumjashka»	-	0.080	0.132	0.252
lamb fat	0.053	0.132	0.238	0.317
desi	0.159	0.360	0.556	0.587
Solubility of fullerene mixture (65% C_{60}, 34% C_{70}, 1% $C_{76\text{-}90}$), g/l; upper value corresponds to solubility of C_{60} from fullerene mixture; lower value corresponds to solubility of C_{70} from fullerene mixture				
Solvent	50^0C	60^0C	70^0C	80^0C
pork fat	0.109	0.437	0.649	0.650
	0.184	0.381	0.571	0.670
bird fat	0.195	0.401	0.533	0.571
	0.191	0.284	0.428	0.495
beef fat	0.126	0.226	0.521	0.551
	0.045	0.190	0.477	0.476
margarine «Rumjashka»	-	0.021	0.135	0.199
	-	0.098	0.193	0.240
lamb fat	0.058	0.282	0.409	0.512
	0.194	0.306	0.460	0.546
desi	0.177	0.261	0.354	0.426
	0.154	0.258	0.309	0.368

"-" denotes that melting point of margarine «Rumjashka» is higher than 50°C.

In Figure 14 the content of the C_{60} fullerene from the mixture C_{60}+C_{70} in a liquid solution after the extraction of 10 mg of fullerene mixture (65 %wt. of C_{60}, 34 %wt. of C_{70}, 1 %wt. of C_{76}+C_{78}+C_{84}+C_{90} ...) by 10 ml of vegetable oil is shown against the temperature.

The Figure 15 shows that at low temperature-extraction (0-40°C) an enrichment of liquid solution in comparison with initial solid fullerene mixture practically do not occur while at high temperatures there is an enrichment of liquid phase by more polarizable C_{70}. The maximal enrichment is observed at extraction by non-purified sunflower oil [26,27].

The polythermal solubility curves of fullerenes C_{60} and C_{70} in natural oils (fats) of animal origin within the temperature range from 50 to 80°C are shown in Figures 16, 17 and Table 6 [27].

Figures 16, 17 show the following [27]:

- The solubility increases monotonously with the temperature increase from 50°C up to 80°C for the both C_{60} and C_{70} fullerenes in all natural oils (fats);
- The C_{70} fullerene solubility in considered natural oils (fats) is always higher than the C_{60} fullerene solubility in the same oils;
- The highest solubility of the C_{60} fullerene at high temperatures is observed in the mutton fat and for the C_{70} fullerene is obsereved in the chicken fat.

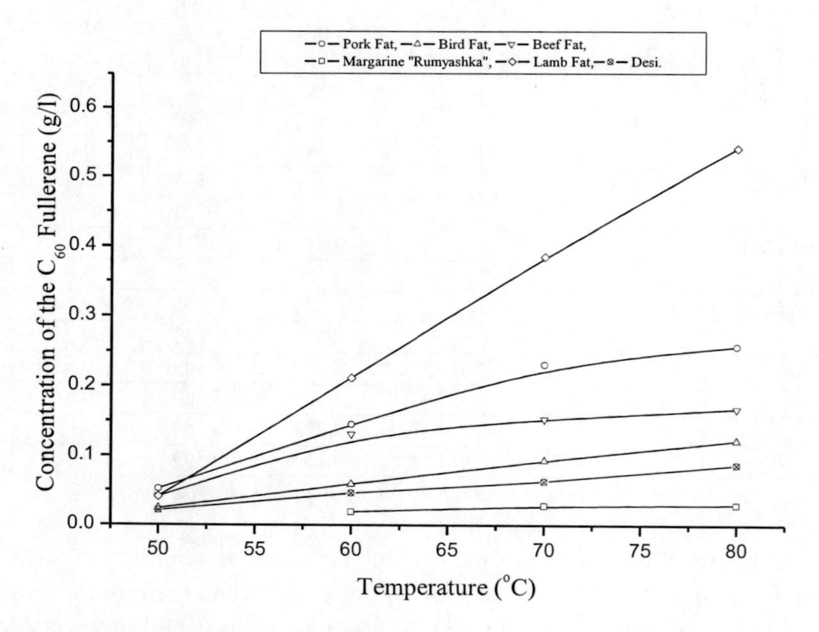

Figure 16. Polythermal solubility of C_{60} in animal fats in the temperature range 50 – 80°C.

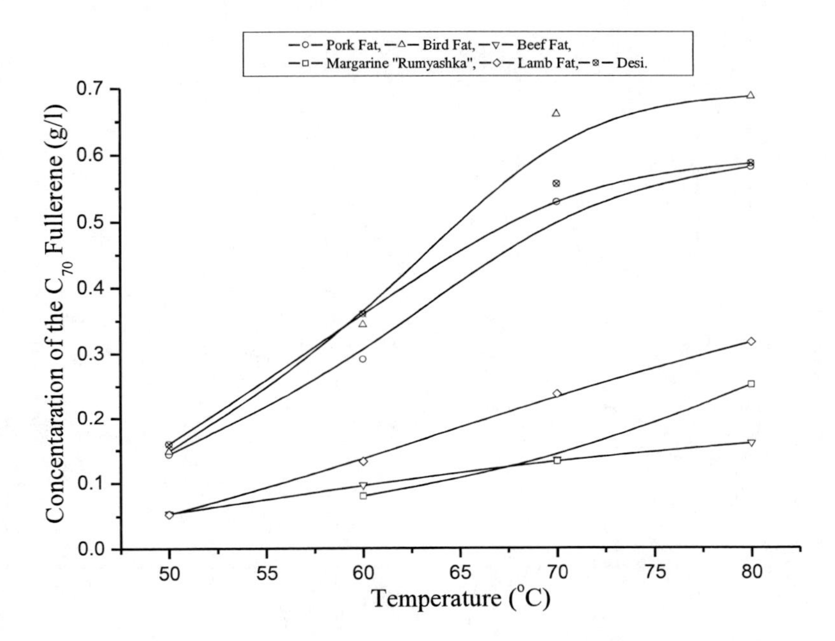

Figure 17. Polythermal solubility of C_{70} in animal fats in the temperature range 50 – 80^{0}C.

The polythermal solubility curves of the fullerenes sum (C_{60}+C_{70}) from fullerene mixture (60 %wt. of C_{60}, 39 %wt. of C_{70}, 1 %wt. of $C_{76\text{-}90}$) in natural oils (fats) are shown in Figure 18 and Table 6 (during the analysis negligible admixtures of the higher $C_{76\text{-}90}$ fullerenes in the abovementioned fullerene mixture were ignored). The solubility of the fullerenes sum (C_{60}+C_{70}) also monotonously increases with the temperature increase from 50 up to 80°C from 0.1-0.4 g l^{-1} up to 0.4-1.3 g l^{-1} depending on the type of natural oil (fat), as shown in Figure 18 [27].

In Figure 19, the content of the C_{60} fullerene from the mixture C_{60}+C_{70} in a liquid solution after the extraction of 10 mg of the fullerene mixture (65 %wt. of C_{60}, 34 %wt. of C_{70}, 1 %wt. of C_{76}+C_{78}+C_{84}+C_{90} ...) by 10 ml of natural oil (fat) is shown against the temperature. The Figure 19 shows that practically at all temperatures there is an enrichment of liquid phase by more polarizable C_{70} in comparison with the composition of initial dissolved solid solution. Unique conversion of relative enrichment of a liquid solution namely depletion of C_{70} (26 %wt. of C_{70} in comparison with composition of initial dissolved solid solution – 34 %wt. of C_{70}) was observed in a unique case of the beef fat at 50°C.

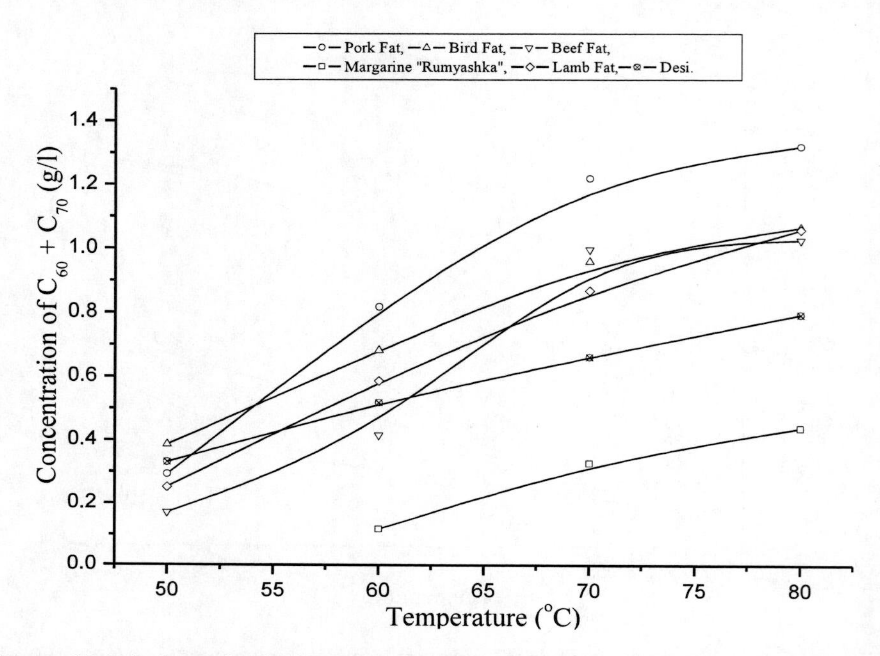

Figure 18. Polythermal solubility of the industrial fullerene mixture (C_{60} – 65 %wt., C_{70} – 34 %wt., C_{76}+C_{78}+C_{84}+C_{90}… - 1 %wt.) in animal fats.

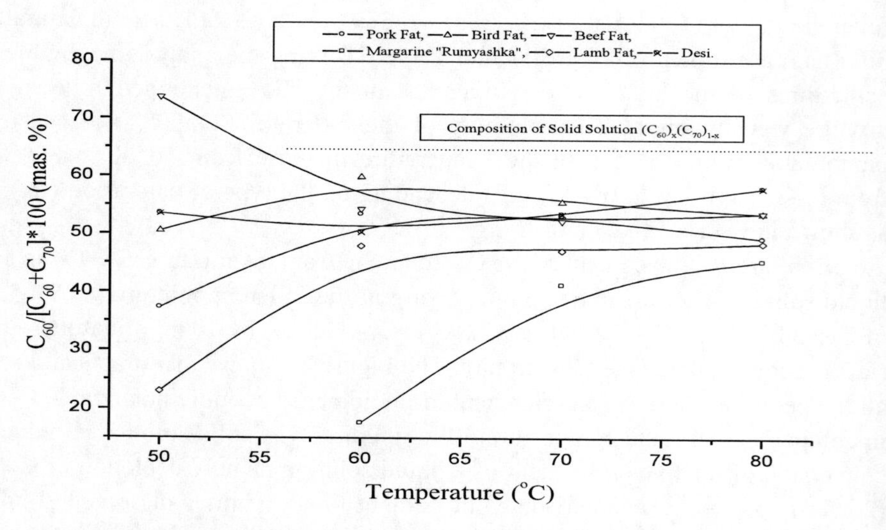

Figure 19. Fullerene content X(C_{60}) in saturated solution of fullerene mixture in animal fats against temperature. The dotted line corresponds to the initial solid solution.

Investigation of solubility of light fullerenes in essential oils is extremely relevant due to the following reasons [28,29]:

- fullerenes solution in natural solvents are stable and absolutely transparent veritable solutions;
- these solutions are absolutely innocuous and biocompatible with animals and human beings ;
- fullerenes solutions in oils and fats possess pronounced antibacterial and “anti-oxidant properties”.
- fullerenes may be also highly effectively used as smell’s fixing agent in the perfumery or cosmetic industry.

It is known that:

- Essential oil of cade can be produced from berries, leafages and from cade wood. Cade is a genus of the evergreen coniferous shrubs and trees of the cypress family (Juniperus),
- Essential oil of carnation is a collective name of essential oils made from flower buds as well as from leafages and tendros of aromatic carnation of the eugenia’s family (Syzygium aromaticum),
- Essential oil of hazel can be obtained from the nucleus of nutwood (Corylus avellana),
- Essential oil of cedar can be manufactured from needle and draws of Atlantic cedar and softwood of the pinaceous family (Pinaceae),
- Essential oil of cypress cam be yielded from the fruits (cypress nuts) as well as in the process of hydrodistillation of leafages and branches of cypress (Cupressaceae),
- Essential oil of lavender can be made from the floriferous inflorescence of spicate lavender (Lavandula angustifolia Mill),
- Essential oil of coriander can be gotten from the fruits of coriander (Coriandrum sativum).

Thus, all of the essential oils chosen for our experiments are of wood origin. According to literature data the compositions of these essential oils are variable; the average compositions of the essential oils used in the frame of this work are given in Table 7 [53].

Table 7. Average macrocomponent composition of essential oils (main componenets) [53]

Composition of the essential oil of carnation			
component	**%wt.**	**component**	**%wt.**
eugenol	27-33	methyl salicylate	0.5-1.5
benzyl benzoate	35-45	heptakozane	15-19
benzyl salicylate	4-6	phenylethyl alcohol	6-8

Composition of the essential oil of coriander			
component	**%wt.**	**component**	**%wt.**
nonane	0.31	α-terpineol	0.09
α-pinene	2.34	cis-2-nonenal	0.23
camphene	0.18	trans-2-nonenal	0.81
β- pinene	0.20	decanal	6.43
octanal	0.62	geraniol	0.68
p- cymene	0.49	trans-2-decenal	25.86
limonene	0.44	trans-2-decen-1-ol	16.22
γ-terpinene	1.71	decanol	4.02
octanol	0.16	undecanal	0.32
linalool	17.77	2-undecenal	1.74
camphor	0.93	trans-2-undecen-1-ol	0.33
dodecanal	0.88	undecanol	0.10
cis-2-dodecenal	0.13	geranyl acetate	0.39
trans-2-dodecenal	7.87	4-dodecen-1-al	0.22
trans-2-dodecen-1-ol	0.71	tridecanal	0.34
2-tridecenal	0.31	2-tetradecenal	5.69
2-hexadecenal	0.32	2-octadecenal	0.28

Composition of the essential oil of cypress			
component	**%wt.**		
	Algeria	**France**	**Argentine**
α- pinene	20.4	40.9	44.5
β- pinene	2.9	0.8	2.0
myrcene	1.3	2.7	0.0
Δ^3-carene	21.5	15.2	30.4
limonene	6.3	2.5	0.0
linalool	0.1	0.8	0.0

Composition of the essential oil of cypress			
component	**%wt.**		
	Algeria	**France**	**Argentine**
borneol	1.0	1.0	0.0
terpinene-4-ol	0.6	1.9	0.0
α-terpineol	1.2	1.4	0.6
terpinene-4-il acetate	2.1	1.2	0.0
α-terpynil acetate	7.0	4.3	0.0
β-elemene	0.0	0.0	1.6
β-cariofillen	0.0	0.0	0.4
γ-bisabolen	0.0	0.0	3.9
α-cedrene	0.4	0.0	0.0

Composition of the essential oil of cedar			
component	**%wt.**	**component**	**%wt.**
β- himachalene	49.03	isoleden	0.64
6-methoxy-acetonaphthone	1.17	longifolen	0.68
δ-cadinene	1.37	α-cedrene	0.95
8- methoxy-acetonaphthone	0.98	himachalane-2,4-diene	0.64
α-kalakoren	1.18	tuopsene	0.39
4-oxy-2-methyl-6-pentylbenzofuran	0.50	α- himachalene	20.26
himachalene oxide	1.00	γ- himachalene	10.75
epi-kubenol	0.62	ar-tumeron	0.34
deodaron	1.65	β-atlanton	0.66
α-atlantol	1.80		

Composition of the essential oil of lavender			
component	**%wt.**	**component**	**%wt.**
linalyl acetate	25-45	camphor	0.0-0.5
1,8-cineol	0-2	octanon-3	0-2
limonene	0.0-0.5	cis-β- ocimene	4-10
trans-β-ocimene	2-6	linalool	25-38
lavandulol	0.0-0.3	terpinene-4-ol	2-6
α-terpineol	0-1	lavandulyl acetate	0-2

Table 7. (Continued)

Composition of the essential oil of hazel-nut			
Component (fatty acid in triglycerides)	**%wt.**	**Component (fatty acid in triglycerides)**	**%wt.**
oleic acid	83-85	stearic acid	2-3
linoleic acid	6-8	palmitoleic acid	0.2-0.4
palmitic acid	5-6	linolenoic acid	0.1
$\alpha-, \beta-, \sigma-$ tocopheroles	0.03-0.05		
Composition of the essential oil of cade			
component	**%wt.**	**component**	**%wt.**
α- pinene	15-25	myrcene	4
sabinene	40-60	terpinolene	2
limonene	2	α-phellandrene	2
p-cymene	4	Hermagren, terpene-4-ol, γ-terpinen, α-tuyen, α-gumulen, β-elemene, β-pinene etc. Up to 150 components	до 20-30

* Composition of the essential oil of cade (wood and leafage) corresponds to composition of essential oil of cade (presented in Table 4) with account of dilution by α -pinene in various ratio ($1/2 \div 1/10$).

The polythermal solubility curves of fullerenes C_{60} and C_{70} in some essential oils spanning temperatures from 20 to 80°C are shown in Figures 20, 21 [28,29].

The same data for the fullerenes sum (C_{60}+C_{70}) from the fullerene mixture (60 %wt. of C_{60}, 39 %wt. of C_{70}, 1 %wt. of C_{76-90}) at the same temperature range are shown in Figure 22 (during analysis, negligible admixtures of the higher C_{76-90} fullerenes in given fullerene mixtures were ignored). Figure 22 shows the following:

The solubility of the fullerenes sum (C_{60}+C_{70}) monotonously increases with increase of temperature from 20 up to 80°C from 1.3 ÷ 9.4 g l^{-1} up to 4 ÷ 34 g l^{-1} depending on the type of essential oil (with an exception of solubility fullerene sum in essential oil of cade, in this case the solubility curve reaches maximum at 50°C). Figures 20, 21 show the analogous case of an increase in the value of solubility following an increase of temperature for individual light fullerenes – essential oils systems[28,29].

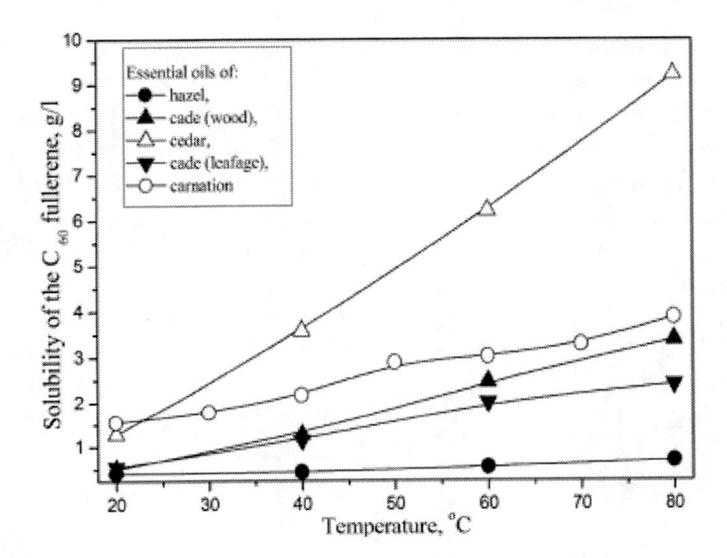

Figure 20. Polythermal solubility of individual C_{60} in some essential oils in the temperature range 20 – 80^{0}C.

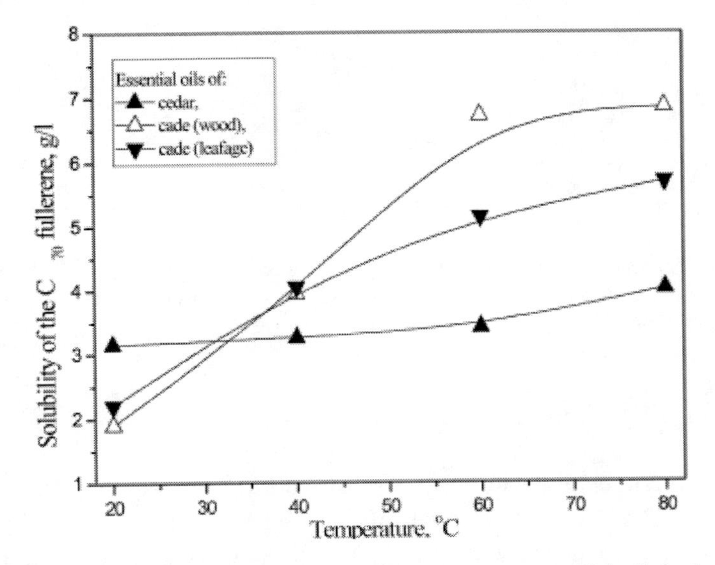

Figure 21. Polythermal solubility of individual C_{70} in some essential oils in the temperature range 20 –80^{0}C.

As is shown in Figure 22, the solubility of the fullerene mixture depending on the type of an essential oil forms the following series characterized by an increase of solubility resulting from an increase of temperature: essential oil of coriander < essential oil of coriander < essential oil of cypress ≈ essential oil of hazel < essential oil of carnation ≈ essential oil of cade (leafage) < essential oil of cade (wood) << essential oil of cedar. In case of temperature dependences of solubility of the individual light fullerenes (C_{60} and C_{70}) another sequence was observed (see Figures 20, 21). The latter fact is caused by another (competitive) type of dissolution of the fullerene's solid solution $(C_{60})_x(C_{70})_{1-x}$ connected with re-iterated processes of dissolution – precipitation of the fullerene components.

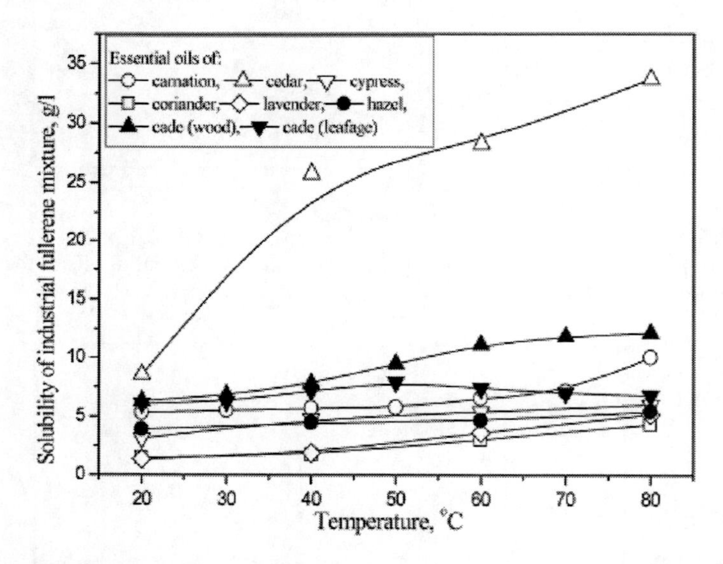

Figure 22. Polythermal solubility of industrial fullerene mixture (C_{60} – 60 mass %, C_{70} – 39 mass %, C_{76}+C_{78}+C_{84}+C_{90}… - 1 mass %) in in some essential oils in the temperature range 20 – 80^{0}C.

2.2. Solubility of Fullerenol in Water

The experimental data on densities of the fullerenol solutions is presented on Figure 23.1 [35-37]. The density of the fullerenol saturated solution increases monotonously with increasing of temperature, the shape of the temperature dependence of density is rather complex, the sigmoid curve was obtained (Figure 23.2.).

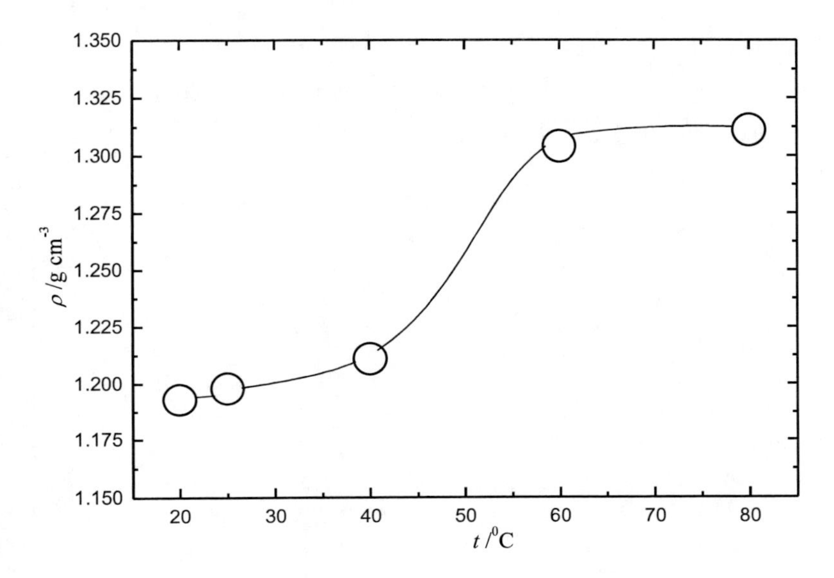

Figure 23.1.

Such type of dependence is influenced by the following factors:

- decreasing of water density and the fullerenol density in any aggregative state (there are no literature data on the fullerenol density, nevertheless the fact of the negative value of the $\partial\rho_{fullerenole} / \partial T$ derivative is also practically assured),
- increasing of the solubility values of the more dense fullerenol in the less dense water in the scale of mass concentrations of the saturated solutions,
- weakening of the physical, in particular Van-der-Vaals forces and simultaneous increasing of the chemical interactions between solvent and solute molecules. Thus, the complex type of the temperature dependence of density is no marvel.

Figure 23.2 illustrates the monotonous increasing of the fullerenol solubility with increasing of temperature; the shape of the $S(T)$ curve is sigmoid. Such form of the temperature dependence of solubility on the branch of crystallization of the individual fullerene (or solid solvated fullerene) is not unusual. For example the temperature dependences of solubility in binary C_{70} + o-xylene system or the branches of crystallization of the sesqui-solvated fullerenes in C_{60} + $\alpha - cloronaphthalene$, C_{60} + $\alpha - bromonaphthalene$ systems are characterized by sigmoid shape of the polytherms of solubility [6,25-28,33]

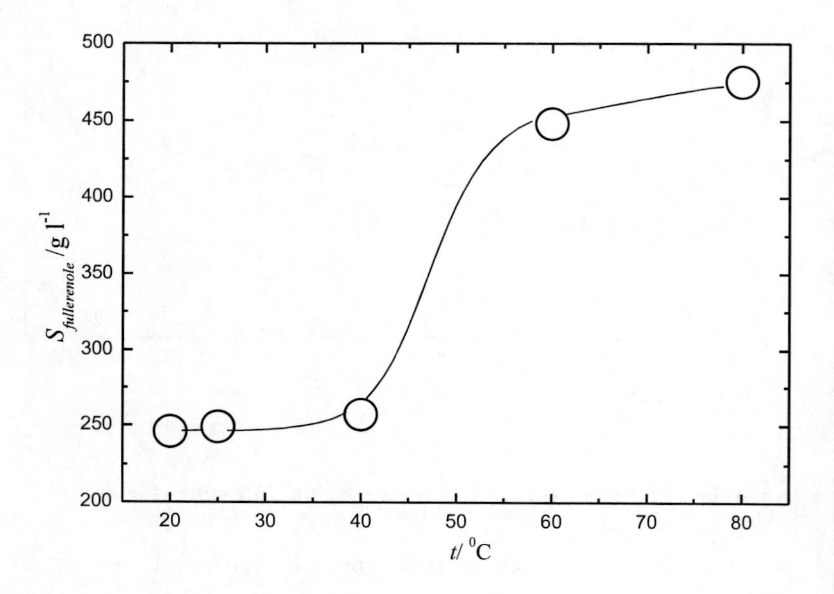

Figure 23.2.

Figure 23. Temperature dependences of density of the fullerenol in water in the temperature range from (20 to 60) °C (23.1). Temperature dependence of the fullerenol solubility in water in the volume concentrations (23.2).

Finally, we can state that the fullerenol solubility is extremely high for the light fullerenes and the light fullerenes derivatives. Practically similar values in the mass concentration and mass fraction scales correspond to solubility of such well-soluble salt as sodium chlorite. Moreover, the fullerenol solubility in water increases greatly with increasing of temperature higher than 50 ^{0}C and exceeds the sodium chloride solubility in the same concentration scales.

Conclusion

In the presented chapter the data on temperature dependences of solubility of individual light fullerenes, industrial fullerene mixture and hydroxyl derivative of C_{60} fullerene are generalized and analyzed. The separate part of the chapter is devoted to analytical methods of determination of fullerenes and fullerenol concentrations in liquid solution (spectophotometery, liquid chromatography) as well as to methods of determination of solid solvates of individual light fullerenes, fullerenol and solvated solid solutions compositions (gravimetric and thermogravimetric analysis). The characteristics of the fullerenes and fullerenol dissolution processes and potential ways of application of the experimental data on fullerenes and fullerenol solubility in biocompatible solvents are discussed in the part devoted to solubility in different types of biocompatible solvents.

REFERENCES

[1] Piotrovskij, L. B., Kiselev, O. I. *Fullerenes in biology*. Saint-Petersburg: Rostok; 2006.

[2] Sidorov, L. N.; Yurovskaya, M. A. *Fullerenes.* Moscow: Ekzamen; 2005.

[3] Hugh, A. *The most beautiful molecule.* New York: JOHN WILEYand SONS; 1995.

[4] Smit V., Bochkov V., Kejpl R. *Organicheskii sintez. Nauka i iskusstvo.* Moscow: Mir; 2001.

[5] Semenov, K. N., Arapov, O. V., Charykov N. A. 2008. The solubility of fullerenes in n-alkanols-1. *Russ. J. Phys. Chem., 82,* 1318-1326.

[6] Heyman, D. 1996. Solubility of C_{60} in alcohols and alkanes. *Carbon*, *34*, 627-631.

[7] Beck, M. T., Mandi, G. 1997. Solubility of C_{60}. *Fullerene Sci. Technol.*, *5*, 291-310.

[8] Korobov, M. V., Smith, A. L. (2000) Solubility of Fullerenes. In K. M. Kadish, R. S. Ruoff (Eds.), *Physics, and Technology* (53-59). New York: A JOHN WIELY and SONS.

[9] Semenov, K. N., Charykov, N. A., Keskinov, V. A., Piartman, A. K., Blokhin, A. A., Kopyrin,

[10] A. 2009. Solubility of light fullerenes in organic solvents. *J. Chem. Eng. Data., 55*, 13-36.

[11] Heymann, D. 1996. Solubility of Fullerenes C_{60} and C_{70} in Seven Normal Alcohols and Their Deduced Solubility in Water. *Fullerene Sci. Technol.*, *4*, 509-515.

[12] Heymann, D. 1996. Solubility of fullerenes C_{60} and C_{70} in water. *Lunar and Planetary Science.*, *27*, 543-544.

[13] Li, J., Takeuchi, A., Ozawa, M., Li, X., Saigo, K., Kitazawa, K. 1993. C_{60} fullerole formation catalyzed by quaternary ammonium hydroxide. *J. Chem. Soc. Chem. Commun.*, *22,* 1784-1785.

[14] Chiang, L. Y., Bhonsle, J. B., Wang, L., Shu, S. R., Chang, T. M., Hwu, R. J. 1996. Efficient one-flask synthesis of water-soluble [60] fullerenols. *Tetrahedron*, *52*, 4963-4672.

[15] Chiang, L. Y., Upasani, R. B., Swirczewski, J. W. 1992. Versatile nitronium chemistry for C_{60} fullerene functionalization. *J. Am. Chem. Soc.*, *114*, 10154-10157.

[16] Meier, M. S., Kiegiel, J. Preparation and characterization of the fullerene diols 1,2-$C_{60}(OH)_2$, 1,2-$C_{70}(OH)_2$, and 5,6-$C_{70}(OH)_2$. *Org. Lett., 3*, 1717-1719.

[17] Szymanska, L., Radecka, H., Radecki, J., Kikut-Ligaj, D. 2001. Fullerene modified supported lipid membrane as sensitive element of sensor for odorants. *Biosens. Bioelectron.*, *16*, 911-915.

[18] Mirkov, S. M., Djordjevic, A. N., Andric, N. L., Andric, S. A., Kostic, T. S., Bogdanovic, G. M., Vojinovic-Miloradov, M. B., Kovacevic, R. Z. 2004. Nitric oxidescavenging activity of polyhydroxylated fullerenol, $C_{60}OH_{24}$. *Nitric Oxide*, *11*, 201-207.

[19] Kokubo, K., Matsubayashi, K., Tategaki, H., Takada, H., Oshima, T. 2008. Facile Synthesis of Highly Water-Soluble Fullerenes More Than Half-Covered by Hydroxyl Groups. *ACS Nano.*, *2*, 327–333.

[20] Yang, J. M., He, W., Ping, W. 2004. Efficient and convenient preparation of water-soluble fullerenol. *Chinese.J.Chem.*, *22*, *9,* 1008-1011.

[21] Sheng, W., Ping, H., Jian-Min, Z., Hu, J., Shi-Zheng, Z. 2005. Novel and Efficient Synthesis of Water-Soluble [60] Fullerenol by Solvent-Free Reaction. *Synthetic Communications, 35, 13,* 1803 – 1808.

[22] Long, Y. United States Patent 5,648,523. Chiang, July 15, 1997.

[23] Semenov, K. N., Charykov, N. A., Letenko, D. G., Nikitin, V.A., Postnov, V. N., Krokhina, O. A. 2009. Synthesis and identification of fullerenol. *Vestnik of Saint-Petersburg State University.*, *4*, 79-86.

[24] Pinteala, M., Dascalu, A., Ungurenasu, C. 2009. Binding fullerenol $C_{60}(OH)_{24}$ to dsDNA. *Int. J. Nanomedicine.*, 4, 193–199.

[25] Cataldo F., Da Ros T. Medicinal chemistry and pharmacological potential of fullerenes and carbon nanotubes. Netherlands: Springer; 2008.

[26] Semenov, K. N., Charykov, N. A., Namazbaev, V. I., Arapov, O. V., Pavlovetz, V. V., Keskinov, V. A., Pyartman, A. K., Strogonova, E. N., Saf'yannikov, N. M. 2009. The solubility of light fullerenes in fats (oils) of animal origin. *Vestnik of Saint-Petersburg State University.*, *3*, 80-87.

[27] Semenov, K. N., Charykov, N. A., Namazbaev, V. I., Arapov, O. V., Pavlovetz, V. V., Keskinov, V. A., Pyartman, A. K. 2009. The solubility of light fullerenes in natural oils. *Russ. J. General Chemistry, 79, 8,* 1323-1330.

[28] Semenov, K. N., Charykov, N. A., Arapov, O. V. 2009. Temperature dependence of solubility of light fullerenes in natural oils and animal fats. *Fullerenes, Nanotubes and Carbon Nanostructures, 17*, 230-248.

[29] Semenov, K. N., Charykov, N. A., Arapov, O. V., Proskurina, O. V., Tarasov, A. O., Strogonova, E. N., Saf'jannikov, N. M. 2010. Solubility of light fullerenes in some natural and essential oils. *Khimija rastitelnogo sirja, 2,* 147-152.

[30] Semenov, K. N., Arapov, O. V., Charykov, N. A., Strogonova, E. N., Saf'yannikov, N. M. 2009. Solubility of light fullerenes in the essential oil of carnation. *Vestnik of Saint-Petersburg State University.*, *1*, 140-144.

[31] Rajhard, K. Rastvoriteli I effekti sredi v organicheskoj khimii. Moscow: Mir; 1991.

[32] Pentin, Yu. A., Vilkov, L. V. Fizicheskie metodi issledovanija v khimii. Moscow: Mir; 2003.

[33] Sajdov, G., Sverdlova, O. Osnovi molekularnoj spektroskopii. Saint-Petersburg: NPO «Professional»; 2006.

[34] Cataldo, F. (2008) Solubility of fullerenes in fatty acids esters: a new way to deliver in vivo fullerenes. Theoretical calculations and experimental results. In F. Cataldo, T. Da Ros (Eds.), *Medical chemistry and pharmacological potential af fullerenes and carbon nanotubes.* Berlin: Springer.

[35] Cataldo, F., Braun, T. 2007. The solubility of C_{60} fullerene in long chain fatty acids esters. *Fullerenes, Nanotubes, and Carbon Nanostructures, 15,* 331-339.

[36] Pogorelij, P.A., Berezin, A.B., Majers, F.E., Roginskij, K.M., Slita, A.V., Kiselev, O.I., Aleksandrov, S.N., Zarubaev, V.V. (27.09.2006) The method of obtaining of fullerene-containing emulsion. Patent RF N 2284293.

[37] Semenov, K. N., Charykov, N. A., Letenko, D. G., Nikitin V. A., Matuzenko, M. Yu., Keskinov, V. A., Postnov, V. N. 2010. Synthesis

and identification of fullerenol. *Peterburgskii jurnal elektroniki, 1*, 41-54.

[38] Semenov, K. N., Charykov, N. A., Keskinov, V. A. 2011. The fullerenol synthesis and identification. Properties of the fullerenol water solutions. *J. Chem. Eng. Data.*, ASAP.

[39] Mischenko, K. P., Ravdel, A. A., Ponomareva, A. M. Practical works on physical chemistry. Leningrad: Khimia; 1982.

[40] Herbst, M. H., Dias, G. H. M., Magalhaes, J. G., Tôrres, R. B., Volpeet, P. L. O. 2005. Enthalpy of solution of fullerene[60] in some aromatic solvents. *J. Mol. Liq., 118*, 9-13.

[41] Shahporonov, M. I. Introduction to molecular theory of solutions. Moscow: Gosudarstvennoe izdatelstvo tehnico-teoreticheskoi literature; 1956.

[42] Ruoff, R. S., Tse, D., Malhorta, R., Lorents, D. C. 1997. Solubility of fullerene C_{60} in a variety of solvents. *J. Phys. Chem., 97*, 3379-3383.

[43] Sivarman, N., Dhamodaran, R., Kallippan, I., Srinivassan, T. G., Vasudeva, P. R., Mathews, C. K. 1992. Solubility of C_{60} in Organic Solvents. *J. Org. Chem. 57*, 6077-6079.

[44] Hansen, C.M., Smith, A.L. 2004. Using Hansen solubility parameters to correlate solubility of C_{60} fullerene in organic solvents and in polymers. *Carbon 42, 8-9,* 1591-1597.

[45] Prausnitz, J., Lichtenthaler, R., Azevedo, E. *Molecular thermodynamics of fluid-phase equilibria*. Englewood Cliffs, NJ: Prentice-Hall; 1986.

[46] Hildebrand, J. H., Scott, R. L. *Regular Solutions*. New York: Prentice–Hall; 1962.

[47] Guggenheim, E. A. *Mixtures.* Oxford: Clarendon Press; 1952.

[48] Dean, J. A. Lange's Handbook of Chemistry. New York: McGRAW-HILL; 1999.

[49] Plugin, A.I., Pogorelij, P.A., Burangulov, N.I., Agafonov, G.I., Slita, A.V., Hubatullin, V.L. (02.10.2001) The method of obtaining of fullerene solutions. Patent RF N 2198136.

[50] Burangulov, N.I., D'jachuk, G.I., Zgonnik, P.V., Mil'rud, E.M., Pogorelij, P.A., Krilova, L.A., Hubatullin, V.L. (10.11.2003) Fullerene-containing cosmetic. Patent RF N 2002106142.

[51] Pogorelij, P.A., Berezin, A.B., Majers, F.E., Roginskij, K.M., Slita, A.V., Kiselev, O.I., Aleksandrov, S.N., Zarubaev, V.V. (10.09.2006) The method of obtaining of fullerene solutions. Patent RF N 2283273.

[52] Berezin, A.B., Pogorelij, P.A., Pogorelij, Yu.P., Majers, F.E., Roginskij, K.M., Slita, A.V., Kiselev, O.I., Aleksandrov, S.N., Zarubaev, V.V.

(10.02.2006) *The method of obtaining of fullerene solutions*. Patent RF N 2004126663.

[53] Hanahan, D. J. *Handbook of Lipids Research*. New York: Plenum Press; 1978.

[54] Guenther, E. *The Essential Oils*. Van Nostrand Company INC: Toronto-New York-London; 1948-1952.

In: Grapes ISBN 978-1-61470-950-3
Editors: R. P. Murphy et al., pp. 51-78

Chapter 2

CHARACTERIZATION OF WINE INDUSTRY RESIDUE AND ITS APPLICATION IN FOODS

Carmen J. Contreras-Castillo, Severino Matias de Alencar, Ligianne Din Shirahigue, Miriam Mabel Selani and Priscilla Siqueira Melo

Department of Agri-Food industry, Food and Nutrition
"Luiz de Queiroz" College of Agriculture
University of São Paulo (USP)
Piracicaba/SP, Brazil

INTRODUCTION

Grape is the second most abundant fruit crop on the planet, following only by orange. According to the Food and Agriculture Organization (FAO), the world production of grapes was approximately 67 million tons. The world wine production reached nearly 27 million tons in 2008, indicating that 40% of the grape production was destined for wine production [1]. Approximately 20% of wine production is represented by pomace (peel and seed), with 5.4 million of tons of wine residues produced every year. These materials include biodegradable residue and their disposal creates serious environmental problems.

Environmental issues have caused increasing interest and concern among all who are involved with agro-industrial activity because the rate of residue generated in these activities is greater than the rate of degradation. Thus, the need to reduce, reuse or recycle residue to recover energy and conserve natural resources is increasing [2].

Peels, seeds and stems are some of the residues resulting from wine processing. Grape seeds contain approximately 40% fiber, 16% oil, 11% protein, and 7% complex phenolic compounds (tannins), sugars, and mineral salts. Grape seeds are rich in essential oil, which has a high aggregate value because it is used in the chemical, pharmaceutical and cosmetics industries. Grape peels are source of proanthocyanidins and anthocyanins, which are inhibitors of lipid peroxidation, and grape peels are also source of natural dyes with natural antioxidant and antimutagenic properties. Grape stems, in turn, are rich in tannin compounds, which have high potential for pharmaceutical and nutraceutical uses [3].

Currently, wine industry residues are underused as agriculture fertilizers or products to feed animals. However, due to the considerable amount of bioactive components in these residues, a nobler destiny could be achieved by the extraction of bioactive substances, which would result in economic gains and decrease the impact of the disposal of these residues in the environment. In the food industry, these substances have the potential to be used as food ingredients, which would provide the consumer with a more natural product, reduce the impact on the environment and bring value to the residue. Several studies have been conducted to evaluate the feasibility of using wine processing residues as food ingredients in many types of foods, such as meat and meat products, seafood, bakery products and dairy products.

The process of lipid oxidation is considered a major problem in food technology. After microbial deterioration, lipid oxidation is the main process by which there is loss of quality, leading to undesirable flavors, and odors, discoloration, production of potentially toxic substances and loss of nutritional value due to the destruction of vitamins and essential fatty acids [4].

The lipid composition of animal products, such as meat, meat products and seafood, is more susceptible to the occurrence of lipid oxidation due to the presence of polyunsaturated fatty acids. In an attempt to control this process, food manufacturers use antioxidants, which are substances that act by preserving and extending the shelf life of foods containing oxidizable lipids through the retardation of oxidation reactions. According to scientific research, however, synthetic antioxidants have toxic potential [5]. Due to increased consumer demand for natural products, particular attention has been given to the use of natural antioxidants, especially those originating from plants [6] and, recently, agro-industry vegetable residues [7].

In this chapter, the chemical composition of grape processing residues and their application in different food matrices will be discussed. Many studies investigating wine residues have already indicated them as sources of natural antioxidants for the food industry, promoting the wine residues as a valuable industrial raw material.

1. Residues from Grape Processing

Peels, pomace, membranes, vesicles and seeds are some of the residues of fruit agro-industrial processing that are generated in large quantities, underused in animal feed and as fertilizers in agriculture, and, often, they become pollutants to the environment [8].

Currently, viable and economic uses for the inevitable agro-industrial residue are being investigated. Whenever possible, the final residue should constitute raw material for a new process. As fruits and vegetables have various components, such as fiber, vitamins, minerals, phenolic compounds and flavonoids, which have health benefits and can prevent diseases [9], several studies have been developed to evaluate the use of these agro-industrial residues as food ingredients.

Approximately 20% of wine production is represented by pomace (peels and seeds), with 5.4 million of tons of wine residues being produced every year. In addition to pomace, other solid materials are generated during the winemaking process. The main byproducts are produced during the destemming (stems) and pressing (peels, seeds and lees) steps [10] (Figure 1).

Pomace is composed of grape solid parts (peels and seeds) and a small proportion of must or the mixture of must/wine in which it is embedded (Figure 2).

Pomace is the product from the pressing step of the wine mass, obtained from either fresh or fermented grapes [12]. Although a fraction of the pomace is distilled to produce an alcoholic beverage called *grappa* [10], the remaining amount discarded is still large.

Stems (Figure 3) are less valuable materials and difficult to utilize. When separated by appropriate machines, stems are approximately 3.5 to 4.5% of the total weight of processed grape [12]. The composition of stems is characterized by high levels of polyphenols, making them unfit for use as livestock feed [10]. The calorific value of stems ranges from 2.000 to 2.500 calories per kilogram, and its use as fuel is absolutely feasible. However, some studies have demonstrated that stems have a rich phenolic composition, providing the possibility of a better appreciationfor this residue [12].

Source: Adapted from [11].

Figure 1. Residues of wine production. A) Red grape stems. B) White grape stems. C) Red grape pomace. D) White grape pomace. E) Red grape lees. F) White grape lees.

Source: [11].

Figure 2. Pomace generated during grape processing in Brazilian wineries.

Source: [11].

Figure 3. Stems generated during the grape processing in Brazilian wineries.

Wine lees are a heterogeneous mass that is deposited in vessels containing wine after fermentation ("first lees" or "fermentation lees") and during storage. Residues obtained by filtration and/or centrifugation of wine and residues deposited in vessels containing grape must are also considered lees. The amount of lees obtained each year depends on several factors, ranging from the constitution of caste, degree of ripeness and hygienic conditions of the grapes to the climatic factors and adopted winemaking techniques.

Due to the involvement of all of these aspects, it is difficult to establish an accurate yield value for lees. Estimates have suggested an amount around 5% of the total volume of wine [12]. Figure 4 shows the appearance of lees after centrifugation of Brazilian red wine.

The flowchart in Figure 5 shows a simplified scheme of red and white wine production, and it indicates the points of residue generation during the process. Currently there is a growing interest in exploiting residues generated by the wine industry. These residues may be an alternative source for obtaining natural antioxidants, which are considered completely safe in comparison to synthetic antioxidants [13].

Grape pomace represents a rich source of various high-value products, such as ethanol, tartrates, malates, citric acid, grape seed oil, hydrocolloids and dietary fiber. Moreover, grape pomace is characterized by high phenolic contents because of poor extraction during winemaking, making their utilization worthwhile and thus supporting sustainable agricultural production [14].

Source: [11].

Figure 4. Lees of red wine generated during grape processing in Brazilian wineries.

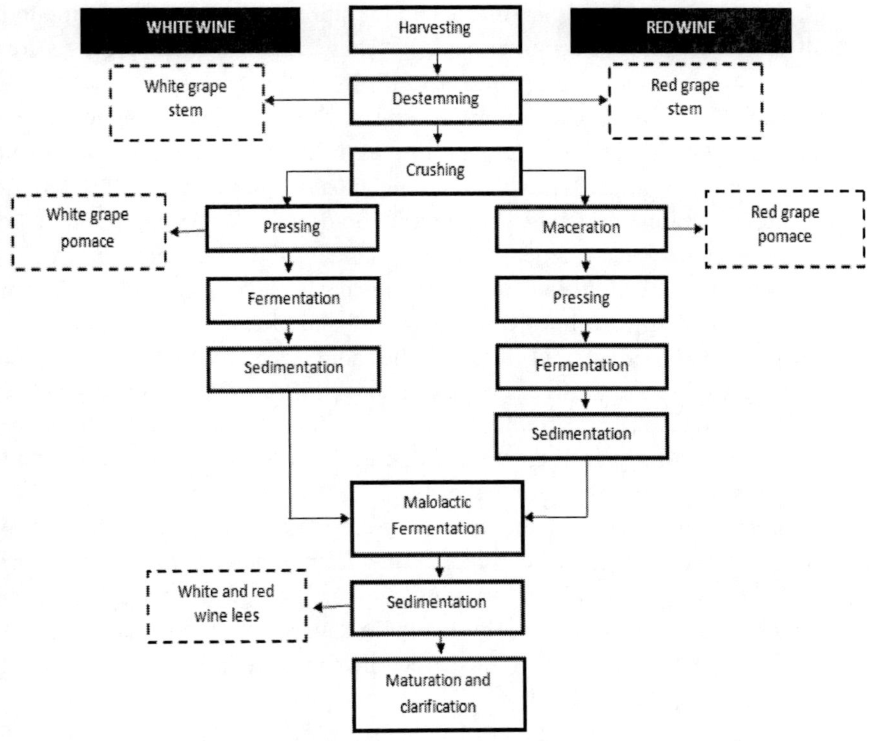

Source: [11].

Figure 5. Flowchart of red and white wine processing with points of residue generation represented by dashed rectangles.

2. Chemical Characterization of Agro-Industrial Grape Residues

Among fruits and vegetables, grapes are considered a major source of phenolic compounds [15]. However, the great diversity among cultivars results in grapes with different characteristics (both in flavor and color), such as in the content and profile of polyphenols [16].

Phenolic compounds are among the most biologically active substances studied in foods. Phenolic compounds have at least one aromatic ring in their structure, and they have one or more hydroxyls as functional groups. These substances are widely found in the plant kingdom, and they are naturally produced by plants in response to environmental changes (infections, injury and ultraviolet radiation). Phenolics include approximately 8000 compounds with different chemical structures, and they have direct influence on sensory parameters such as color and astringency [17-18].

The phenolic compounds in grapes and in their residues (pomace, stems and lees) can be divided into two groups as follows: 1) phenolic acids and related compounds and 2) flavonoids. The most common phenolic acids in grapes include cinnamic acids (coumaric acid, caffeic acid, ferulic acid, chlorogenic acid and neochlorogenic acid) and benzoic acids (p-hydroxybenzoic acid, vanillic acid and gallic acid). Flavonoids include flavan-3-ol (catechin and epicatechin), flavanones, red and blue anthocyanins [19].

Powerful antioxidants, such as gallic acid, catechin, epicatechin and quercetin, have been identified in grape pomace [10, 20-21], grape seeds [22] and grape stems [23]. Currently, the great interest in various bioactive substances, particularly antioxidants, is due to the effects of these compounds on free radicals, providing health benefits.

As juices and wines, also the residues generated during the vinification of grapes have sparked scientific interest due to the wealth of the chemical composition and the significant amount of phenolics generated by wineries, which makes these residues widely available sources of natural antioxidants. Thus, there is an emerging new industry within the wine industry that is taking advantage of these residues [24].

The chemical characterization of the main residues generated in the wine industry (pomace, stem and lees) is discussed in further detail below.

2.1. Grape pomace

Except for water, grape pomace is mainly composed of dietary fiber (Table 1), which is formed by cell wall polysaccharides and lignin. Pectin is the main constituent of cell walls in pomace and fresh grapes, and white grape pomace has higher levels of pectin than red grape pomace. Cellulose is the second type of cell wall polysaccharides that exist in high quantities in grape pomace. Hemicellulose content does not significantly differ between grapes and their byproducts (pomace and stems), suggesting that these wine byproducts are sources of quality dietary fiber [25]. With regard to the lipid content, grape by-products has a significant variation of the lipid content in seeds among five grape varieties (Baladi black, Asbani, Baladi green, Ajloni, and Khudari), and no difference among the grape varieties for other chemical constituents, such as water, ash and protein [26].

Table 1. Chemical composition of several varieties of red and white grapes (g/100 g of fresh weight)

Grape pomaces (varieties)	Moisture	Protein	Lipid	Soluble sugars	Dietary fibers	Ash	References
C. Sauvignon[1]	10.2-61.5	3.1-11.3	0.6-7.3	2.3-5.32	30.4-56.9	2.1-7.58	[1-2]
Royal Rouge[1]	9.98	11.1	6.80	6.03	55.4	6.97	[1]
Callet	55.6	2.7	0.3	2.5	37.2	2.2	[2]
Manto Negro	63.6	3.2	0.7	3.1	27.6	2.1	[2]
Merlot	53.9	3.8	0.5	2.4	37.4	2.1	[2]
Tempranillo	55.7	3.1	0.6	2.0	36.9	2.3	[2]
Syrah	50.2	3.3	0.7	2.1	40.8	3.0	[2]
Chardonnay	63.9	2.7	1.0	6.2	23.5	1.8	[2]
Macabeu	72.2	3.2	0.9	2.1	19.9	1.9	[2]
Parellada	62.8	3.4	0.7	2.2	30.2	1.8	[2]
Premsal B.	69.3	3.1	0.7	2.1	21.9	2.0	[2]

[1]grape pomace powder.
(1) [24].
(2) [25].

The triglyceride fractions of Cabernet Sauvignon and Royal Rouge grape pomace varieties have a high content of polyunsaturated fatty acids (64.4 and 66.2%, respectively) and a high n-6/n-3 ratio (70.7 and 55.1, respectively) [24]. Grape seeds have been investigated for their health benefits, due to high content of polyunsaturated fatty acids. Studies have indicated that these fatty acids aid in the prevention of cardiovascular disease [27] and cancer [28]. The main lipophilic constituents of grape pomace are not commonly reported. However, lupeol, oleanoic acid and β-sitosterol glucoside have been identified in ethanol extracts of grapes. These compounds have known biological activities. For example, lupeol has anti-inflammatory activity [29] inducer of apoptosis [30] and an inhibitor of tumors [31].

There are many differences among data reported for the phenolic content and antioxidant activity of wine residues, mainly due to the differences among grape varieties (Table 2), geographical origin and the extraction method employed in the studies.

These differences complicate comparisons among wine residues. However, high phenolic contents and antioxidant activities of grape pomace, stems and seeds have been reported by several authors [10, 23, 32-33].

Table 2. Phenolic compound contents in pomace from several grape varieties (mg gallic acid equivalents (GAE)/g)

Grape pomaces (varieties)	Phenolic compounds	References
Cabernet Sauvignon	2.50±2.6	(1)[1]
Cabernet Sauvignon	74.75±2.2	(2)[2]
Royal Rouge	4.80±2.8	(1)[1]
Pinot Noir	618±0.8	(3)[3]
Cabernet Franc	469±4.1	(3)[3]
Baco Noir	515±0.4	(3)[3]
Noiret	411±1.1	(3)[3]
Merlot	46.23±1.6	(2)[2]
Bordeaux	63.31±2.4	(2)[2]
Isabel	32.62±0.6	(2)[2]
Muscadine	34.1±1.8	(4)[3]

(1) [24] (2) [34] (3) [35] (4) [20].
[1] values expressed as fresh weight.
[2] values expressed as dry weight.
[3] values expressed per gram of grape pomace extract.

In grapes and wines, the major polyphenols are proanthocyanidins, which are oligomers and polymers of polyhydroxiflavan-3-ol monomeric units that are most commonly linked by acid-labile bonds (4→8 and 4→6). The most common compounds of this class are (+)-catechin and (-)-epicatechin. Grape seeds are rich in these compounds, which are present in greater quantities in the seeds than in the peel and stems. Procyanidins from grape seeds have higher antioxidant activity than vitamin C, and this antioxidant activity is positively correlated with their degree of polymerization [22].

Furthermore, the arrangement and number of hydroxyl groups of a compound can influence its ability to donate electrons and hydrogen. Therefore, this ability and, consequently, the antioxidant activity of the compound is increased by increasing the number of hydroxyl groups on phenol [33]. Epicatechin has a total of five hydroxyl groups that are distributed to create an ortho-dihydroxy structure (catechol in ring B), which contributes to the displacement of electrons and confers high stability to the molecule. Moreover, the ortho-dihydroxy structure is the most active structure in donating hydrogen atoms for free radicals. The other three free hydroxyl groups distributed in rings A and C also participate in the donation of hydrogen and in stabilizing the molecule by the formation of hydrogen bonds and displacement of electrons [17, 33]. Epicatechin is characterized as a substance with exceptional antioxidant power (Figure 6).

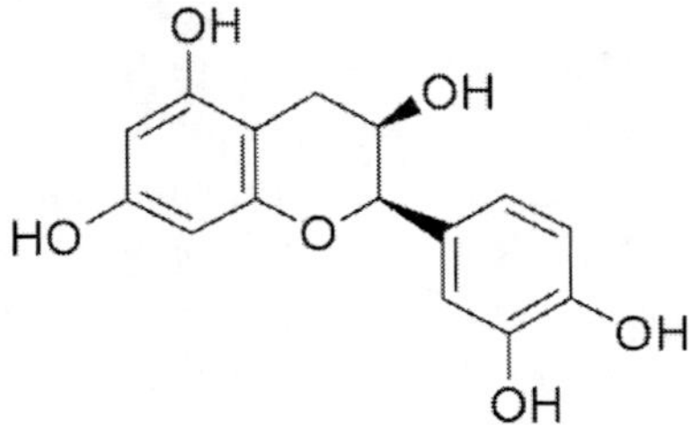

Figure 6. The epicatechin molecule with the A, B and C rings identified.

Eleven flavan-3-ols were identified in grape seeds, with catechin and epicatechin being the most abundant. Therefore, grape seeds and their peels contribute to the contents of catechin, epicatechin and procyanidins in wines and juices. These compounds affect the color, clarity, flavor, astringency, stability and aging characteristics of these beverages [36]. Moreover, they confer nutraceutical features because have high antioxidant potential and contribute to the prevention of cardiovascular [37], inflammatory [38], neurodegenerative diseases [39] and cancer [40].

In Chardonnay grape pomace were isolated and identified 17 polyphenols, including phenolic acids (gallic acid; 3- and 4-b-glucopyranosides of gallic acid; trans-caftaric acid; and cis- and trans-coutaric acids), phenol alcohols (2-hydroxy-5-(2-hydroxyethyl)phenyl-b-glucopyranoside), flavan-3-ols (catechin, epicatechin and procyanidin B1) and flavonoids (quercetin-3-glucoside, quercetin-3-glucuronide, kaempferol 3-glucoside, kaempferol 3-galactoside, eucryphin, and astilbin engeletin) [41]. Table 3 shows the polyphenolic composition of pomace from several grape varieties.

In Muscadine grape pomace were detected ellagic acid, gallic acid, (-)-epigallocatechin, (-)-epicatechin, catechin, myricetin, quercetin, kaempferol and 3,3',4,4'5,5'-hexahydroxystilbene. However, resveratrol was not detected in this grape pomace even though it has been previously identified by others. Fermentation can convert resveratrol into its analog (3,3',4,4'5,5'-hexahydroxystilbene), which explains the high content of this compound in grape pomace [20].

In a chemical study of the methanol extract of Nerello Mascalese grape pomace, were found quercetin-3-O-glucoside and quercetin-3-O-glucuronide to be the most abundant flavonol glycosides and malvidin 3-O-glucoside to be the major anthocyanin in this extract [21].

In addition to the compounds abovementioned, anthocyanins are among the main bioactive constituents of grape pomace. These molecules are phenolic compounds that belong in the class of flavonoids consisting of three phenolic rings with glycosidic substitutions at positions three and five. Anthocyanins are natural pigments that are water soluble, which provide the red, blue, purple and violet colors to various flowers and fruits. Anthocyanins are valued for their high antioxidant activity and health benefits arising from the antioxidant activity, as previously demonstrated [42, 43, 44]. Moreover, anthocyanins have potential as natural colorants and antioxidants for industry and food science [44-45].

Figure 7 shows the generic molecular structure of an anthocyanidin (aglycone form) and the R1 and R2 substitution sites in addition to the functional groups characteristic of six naturally occurring anthocyanidins

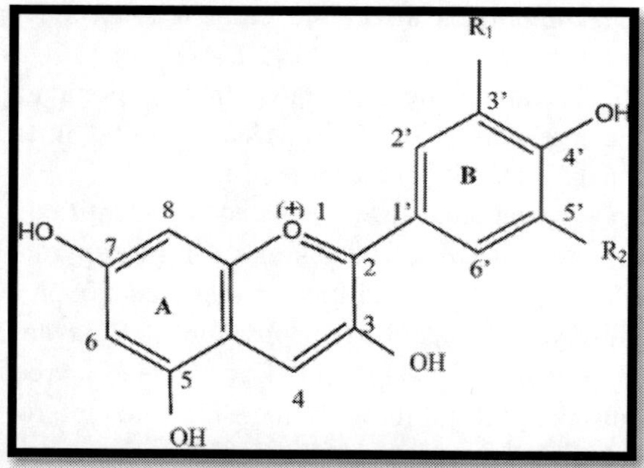

Anthocyanidin	R_1	R_2
Pelargonidin	H	H
Cyanidin	OH	H
Delphinidin	OH	OH
Peonidin	OCH_3	H
Petunidin	OCH_3	OH
Malvidin	OCH_3	OCH_3

Figure 7. Structure of six anthocyanidins (no sugar attached at position 3) of natural occurrence with the A and B aromatic rings and R1 and R2 substitution sites. Adapted from [45].

Table 3. Major phenolic compounds in grape pomace extracts (extract mg/100 g)

Grape pomaces	Phenolic compounds							references
(varieties)	gallic acid	catechin	epicatechin	resveratrol	miricetin	Quercetin	kaempferol	
Pinot Noir	75±0.004	8100±0.12	629±0.004	25±0.001	ND	128±0.004	165±0.002	(1)
C. Franc	100±0.014	9730±0.01	616±0.011	22±0.001	54±0.003	375±0.011	155±0.006	(1)
Baco Noir	120±0.011	13090±0.09	966±0.012	ND	7±0.001	260±0.004	55±0.001	(1)
Noiret	ND	5310±0.10	1021±0.020	44±0.001	363±0.007	289±0.001	ND	(1)
C. Sauvignon	4.59±0.03	150.16±0.91	18.24±0.12	4.02±0.39	ND	13.98±0.66	ND	(2)
Merlot	9.08±0.04	122.29±1.18	25.97±0.57	6.40±0.20	ND	56.65±0.22	15.09±0.12	(2)
Bordeaux	18.68±0.08	111.61±1.79	ND	ND	ND	9.18±1.24	ND	(2)
Isabel	17.49±0.02	94,28±0.08	44.36±0.14	1.18±0.08	ND	25.72±0.35	11.41±0.51	(2)
Muscadine	25±0.01	146±0.21	128±0.03	ND	17±0.01	20±0.0	8±0.01	(3)

(1) [35] (2) [34] (3) [20].

In extracts from Sunbelt grape pomace, was found malvidin-3,5-O-diglucoside as the most abundant anthocyanin followed by peonidin-3-(6-Op-coumaroyl)-5-O-diglucoside. The composition of anthocyanins in grapes is influenced by genetic factors, but the anthocyanin content is determined by maturity, seasonal variations, environment and soil conditions [45].

2.2. Grape Stems

In grapes, the fraction of soluble sugars is the main constituent. However, in stems, dietary fiber is the most abundant fraction, as observed for pomace (Table 4). Cellulose is the main cell wall polysaccharide of stems [25]. Dietary fibers are carbohydrate polymers resistant to hydrolysis by enzymes of the human digestive system.

These fibers help in preventing constipation and obesity, lowering blood cholesterol levels and the risk of cardiovascular disease. The main sources of these fibers in the human diet are cereals, however, grape pomace and stems are also good sources [25].

Table 4. Chemical composition of stems of several red and white grape varieties (g/100 g of fresh weight)

Grape stems (varieties)	Moisture	Protein	Lipid	Soluble sugars	Dietary fibers	Ash	References
Cabernet Sauvignon	65.8	2.0	0.9	3.1	23.9	3.7	[1]
Callet	66.3	2.8	0.7	3.1	25.5	2.4	[1]
Manto Negro	59.7	2.7	0.4	2.9	30.0	2.8	[1]
Merlot	61.5	2.2	0.9	2.9	28.8	4.3	[1]
Tempranillo	56.8	2.1	0.4	2.2	34.8	4.3	[1]
Syrah	64.9	2.4	0.9	3.0	26.3	1.7	[1]
Chardonnay	67.5	2.5	0.7	1,8	26.0	2.8	[1]
Macabeu	72.7	1.8	0.5	3.2	19.5	1.5	[1]
Parellada	76.7	2.6	0.8	2.9	16.5	1.5	[1]
Premsal B.	69.7	2.8	0.7	3.7	21.2	1.8	[1]

(1) [25].

Grape stems are also sources of phenolic compounds with known bioactive activities. Merlot grape stems have lower levels of caftaric and coutaric acids than the levels present in the pulp and peel. However, these stems have significant amounts of quercetin 3-glucuronide and astilbin, which are the main flavanonols in the stems of white and red grape varieties [46]. Table 5 shows the concentration of phenolic compounds in Merlot grape stems.

Table 5. Concentration of phenolic compounds in Merlot grape stems

Compounds	mg/kg stem
Caftaric	40
Coutaric	4.5
Quercetin 3-glucuronide	200
Quercetin 3-glucoside	18
Kaempferol 3-glucoside	Traces
Myricetin 3-glucoside	Traces
Myricetin 3-glucuronide	Traces
Catechin	60
Epicatechin	Traces
Astilbin	35
Engeletin	Traces

Source: [46].

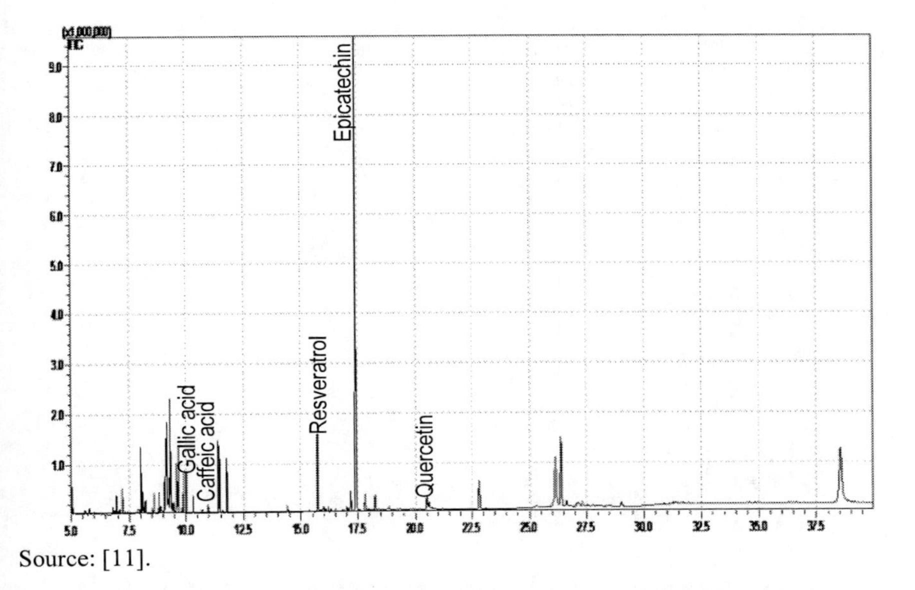

Source: [11].

Figure 8. Gas chromatographic profile produced by gas chromatography with mass spectrometry (GC-MS) of an ethanol extract of Moscato white grape stems.

Among the phenolic compounds identified in Moscato white grape stems, epicatechin is the major compound (Figure 8). Moreover, the resveratrol peak is more abundant in stems than in the pomace of this variety. According to previously presented data, the phenolic composition varies when there is alteration in the grape variety.

Grape stems are greatly generated by the wine industry. Similarly to pomace, grape stems are excellent sources of compounds of interest to the food and pharmaceutical industry, such as dietary fiber and polyphenols.

2.3. Lees

Studies on the chemical composition of wine lees are rare. However, isovanilic acid, syringic acid, p-coumaric acid, gallic acid, ferulic acid, caffeic acid, resveratrol, quercetin and epicatechin have been identified in a hydroalcoholic extract of grape lees using gas chromatography with mass spectrometry (GC-MS) [11]. Figure 9 shows the chromatogram obtained by GC-MS of the extract from red wine lees.

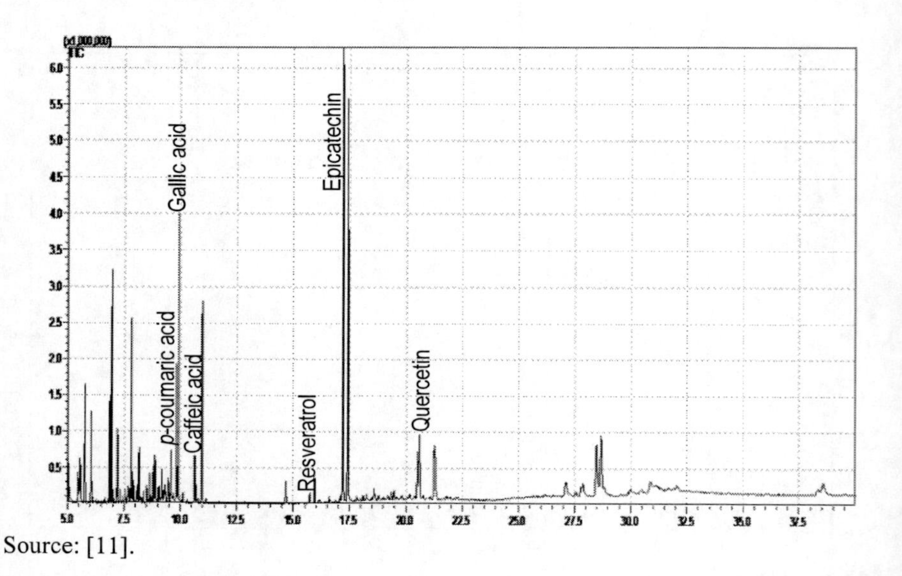

Source: [11].

Figure 9. Gas chromatographic profile produced by gas chromatography with mass spectrometry (GC-MS) of an ethanol extract of red wine lees.

Similar to grape stems and pomace, lees generated from wine processing also have many phenolic compounds. Thus, lees are a natural source of these compounds.

3. Lipid Oxidation Products

After microbial deterioration, lipid oxidation is the main process by which there is loss of quality in food products [47], and is an important factor in determining shelf life [48].

This oxidation process is mainly related to the unsaturated fat and fundamental constituents of the structure responsible for ensuring the functional integrity of biological membranes.

The following factors influence the oxidation of foods: the quantity and availability of oxygen present; the instauration degree of fatty acids and the presence of metals and enzymes. However, it is difficult to assess the effect of a specific factor in the oxidation process because these factors act simultaneously [49].

4. Antioxidants

The term antioxidant can be defined as any substance when present in low concentrations in comparison with other oxidizable substrates that significantly delay or prevent oxidation of that substrate [50]. The addition of antioxidants is a method that can extend the shelf-life, especially for food containing lipids [51].

Antioxidant substances can act in several ways: binding competitively with oxygen; slowing the initiation step and/or interrupting the propagation step for the destruction or binding of free radicals; inhibiting catalyst activity; or stabilizing hydroperoxides [49].

The action of antioxidants on cell membranes may be as follows: 1) to scavenge free radicals, thereby preventing the oxidation process; 2) to inactivate metal ions; 3) to remove reactive oxygen species; 4) to scavenge singlet oxygen molecules; 5) to destroy peroxides and prevent radical formation; and 6) to remove and/or decrease the local oxygen concentration [52-53].

4.1. Mechanism of Action of Antioxidants

Antioxidants may act at different steps of the oxidation process depending on their mode of action. The primary antioxidants react with lipid radicals to produce more stable molecules, and they are known as free radical interceptors. Secondary or preventive antioxidants slow the initiation step by several mechanisms, including inactivation of metals, decomposition of hydroperoxides and oxygen sequestration by synergism.

Primary antioxidants quickly donate a hydrogen atom to a lipid radical. The inhibition of free radicals occurs in two major steps in the chain reaction of lipid oxidation. Primary antioxidants react with peroxil radicals (LOO•), preventing the spread of chain reaction, inhibiting the formation of peroxides, and react with alkoxyl radicals (LO•) to reduce the decomposition of hydroperoxide degradation products [54].

$$LOO\bullet + AH \rightarrow LOOH + A\bullet$$
$$LO\bullet + AH \rightarrow LOH + A\bullet$$

Metal inactivators or chelating agents act as preventive antioxidants by inhibiting the effect of metallic ions, which act as pro-oxidants and catalyze the decomposition of hydroperoxides [54]. Synergism can be expected between substances with different modes of action, so systems of antioxidant compounds have greater activity.

4.2. Antioxidants in Food

Acceptable antioxidants for use in foods should be effective at low concentrations, compatible with substrates, sensory acceptable, nontoxic and not affect the physical properties of foods [55].

Synthetic antioxidants commonly used are butylated hydroxyanisole (BHA), butylatedhydroxytoluene (BHT), tertiary butyl hydroquinone (TBHQ) and propyl gallate (PG).

There are various regulations in different countries to control the amount or application of these antioxidants in foods. These regulations restrict their use due to suspect to be carcinogens [56].

These substances have been widely used to retard lipid oxidation in foods [57]. However, due to potential risks of synthetic antioxidants to human health and the evidence showing the ability of plant extracts to reduce or retard oxidation in foods the application of natural antioxidants has increased in recent years [58,6].

5. Applications of Grape Processing Residues in Foods

In recent years, the use of grape seed extracts has gained importance as a nutritional supplement because of their antioxidant activity [59]. The byproducts obtained from wineries constitute an inexpensive source for the extraction of antioxidant, which can be used as food ingredients in industry. The use of these materials as a source of natural antioxidants has been the object of study in several food matrices, such as meat, seafoods, bakery products and dairy products.

5.1. Meat and Meat Products

The use of wine residues as natural antioxidants has been widely studied in various types of raw materials, such as beef, chicken, pork and turkey, using various matrices, including burgers, meatballs, restructured meats, sausages and marinated cuts.

The addition of Isabel and Niagara grape seed and peel extract as a natural antioxidant retarded lipid oxidation of cooked chicken meat stored under refrigeration for 14 days, with results comparable to the effects of butylhydroxytoluene, a synthetic antioxidant [60].

Using the same grape pomace varieties (Isabel and Niagara), it was demonstrated that grape seed and peel extracts of both varieties were as effective as the synthetic antioxidants BHT and sodium erythorbate in the prevention of lipid oxidation in both raw and cooked chicken products stored under freezing conditions for nine months. Despite the satisfactory effect in retarding lipid oxidation, the addition of natural extracts, especially the Niagara grape extract, interfered with the natural color, flavor and odor of the chicken meat [61].

The antioxidant effects of grape seed extract (0.02%), oleoresin rosemary (0.02%), oregano extract (0.02%), propyl gallate (0.02%), butylhydroxyanisole (BHA) (0.02%) and butylhydroxytoluene (BHT) (0.02%) on lipid oxidation and color stability were evaluated in restructured pork stored under freezing conditions for six months [62]. This study showed that the grape extract and propyl gallate presented higher antioxidant activity than the other antioxidants during storage of the product and that the grape extract did not affect the color parameters of the restructured pork.

The addition of grape seed and green tea extracts on beef patties with low sulfite delayed microbial spoilage, redness loss and lipid oxidation, thus increasing the shelf life of the raw beef patties by three days. Grape seed and green tea extracts also delayed the onset of rancid flavors in cooked patties. Moreover, no anomalous sensory traits were caused by either extract. So, ascorbate, green tea extract and grape seed extract improved the preservative effects of sulfur dioxide on beef patties, especially against meat oxidation [63].

5.2. Seafood

The lipids of fish are rich in omega-3 polyunsaturated fatty acids, especially eicosapentaenoic acid (EPA) and docosahexaenoic acid (DHA), which can reduce levels of triglycerides and blood cholesterol and thereby reduce the risk of cardiovascular disease [64]. Seafood is an excellent source of important minerals, is rich in water soluble vitamin of B complex and fat soluble vitamins, such as A and D [71].

However, lipid oxidation is a major cause of fish deterioration because the fatty acids of aquatic organisms are highly unsaturated. These changes affect the nutritional value and quality, as manifested by changes in odor, color, texture and possible production of toxic compounds, influencing the consumer acceptability of the fish. Processing, milling, and cooking processes, such as salting and smoking, promote the oxidation. The use of antioxidants in food coupled with appropriate packaging and storage under refrigeration and/or freezing can protect against this chemical degradation and, consequently, against the loss of essential nutrients in the lipid fraction.

Recent studies have demonstrated the importance of winemaking byproducts as a rich source of phenolic compounds, especially flavonoids [65, 66]. Moreover, grape polyphenols are effective in retarding lipid oxidation in fish-based products during frozen storage [67]. The antioxidant activity of flavonoids from grape (*Vitis vinifera*) byproducts from the wine industry was evaluated in systems containing fish oil and Mackerel muscle (*Scomber scombrus*), and it was observed that flavonoid oligomers have high potential to inhibit oxidation in emulsions with fish oil and in frozen fish muscle. Other study also evaluated the ability of grape extract (*Vitis vinifera*) and its purified fractions in the preservation of meat fish species with high concentrations of polyunsaturated fatty acids [68].

5.3. Bakery Products

The addition of fruit byproducts (grape seeds, orange peels and tomato pomace) to rice-based extruded foods demonstrate that fruit byproducts can be an alternative source of natural antioxidants in the enrichment process of products [69].

Breads made with grape seed extract have showed an increase in antioxidant activity and a decrease in the levels of carboxymethyl lysine, which is a potentially toxic compound formed via the Maillard reaction. Furthermore, among all of the attributes evaluated, only the color of the bread was affected by the use of grape seed extract [70].

A previous study evaluated the application of grape pomace extract as a bread improver. The study demonstrated that the addition of grape pomace extracts caused a slight drop in farinographic stability and an increase in the strength of wheat flour when compared to flour without additives. In addition, breads prepared with grape pomace extracts showed higher specific volume when compared to bread without additives, resulting in breads with higher growth [71].

5.4. Dairy Products

The antibacterial effect against *Listeria monocytogenes*, *Staphylococcus aureus* and enteric *Salmonella* and the antioxidant effect of five natural extracts (cinnamon, oregano, clove, pomegranate peels and grape seeds) were investigated in cheddar cheese. The study demonstrated that all five plant extracts were effective against the three foodborne pathogens. Moreover, thetreatments with these extracts increased the stability of the cheese against lipid oxidation [72].

The addition of lees from wine production was evaluated in ice cream, and positive effects were found, as the increase in melting rate, fat destabilization and the decrease of freezable water. Thus, the addition of grape wine lees significantly increased the antioxidant activity of ice cream, and the antioxidants of the residue presented stability during the process of ice cream making. Therefore, grape wine lees have potential to be used as a value-added ingredient in the ice cream industry to enhance antioxidant activity [73].

Conclusion

Wine residues have phenolic compounds, especially flavan-3-ols (catechin and epicatechin), which have high antioxidant, anti-inflammatory and antiproliferative effects. Thus, wine residues are promising sources of bioactive compounds for the pharmaceutical and food industries.

According to the literature, grape residue extracts have antioxidant effects in several types of food, showing great potential for use by the food industry and in the development of functional foods.

References

[1] Food and Agriculture Organization - FAO. (2008). FAOSTAT. Available in: <http://faostat.fao.org>.

[2] Straus, E.L. and Menezes, L.V.T. (1993). Minimização de resíduos. Anais do Congresso brasileiro de engenharia sanitária e ambiental, 17, 212-225.

[3] Murga, R., Ruiz, R., Beltran, S. and Cabezas, J.L. (2000). Extraction of natural complex phenols and tannins from grape seeds by using supercritical mixtures of carbon dioxide and alcohol. *Journal of Agricultural and Food Chemistry,* 48, 3408-3412.

[4] Gray, J.L., Gomaa, E.A. and Buckley, D.J. (1996). Oxidative quality and shelf life of meats. *Meat Science*, 43, S111-S123.

[5] Nantitanon, W., Yotsawimonwat, S. and Okonogi, S. (2010). Factors influencing antioxidant activities and total phenolic content of guava leaf extract. *LWT - Food Science and Technology*, 43, 1095-1103.

[6] Jayaprakasha, G.K. and Jaganmohan Rao, L. (2000). Phenolic constituents from lichen Parmotrema stuppeum (Nyl.) Hale and their antioxidant activity. *Zeitschrift für Naturforschung*, 55, 1018-1022.

[7] Conde, E., Moure, A., Domínguez, H. and Parajó, J.C. (2011). Production of antioxidants by non-isothermal autohydrolysis of lignocellulosic wastes. *LWT - Food Science and Technology*, 44, 436-442.

[8] Thassitou, P. K. and Arvanitoyannis, I. S. (2001). Bioremediation: a novel approach to food waste management. *Trends in Food Science and Technology*, 12,185-196.

[9] Aruoma, O. I. Nutrition and health aspects of free radicals and antioxidants. (1994). *Food Chemistry and Toxicology*, 32, 671-683.

[10] Ruberto, G., Renda, A., Daquino, C., Amico, V., Spatafora, C., Tringali, C. and De Tommasi, N. (2007). Polyphenol constituents and antioxidant activity of grape pomace extracts from five Sicilian red grape cultivars. *Food Chemistry*, 100, 203-210.

[11] Melo, P.S. (2010). Composição química e atividade biológica de resíduos agroindustriais. Dissertação (Mestrado em Ciência e Tecnologia de Alimentos) – Escola Superior de Agricultura "Luiz de Queiroz", Universidade de São Paulo, Piracicaba, São Paulo, Brazil.

[12] Silva, L.M.L.R. (2003). *Caracterização dos subprodutos da vinificação.* Millenium, 28, 123-133.

[13] Arvanitoyannis, I.S., Ladas, D. and Mavromatis, A. (2006). Potential uses and applications of treated wine waste: a review. *International Journal of Food Science and Technology*, 41, 475-487.

[14] Kammerer, D., Claus, A., Carle, R. and Schieber, A. (2004). Polyphenol screening of pomace from red and white grape varieties (Vitis vinifera L.) by HPLC-DAD-MS/MS. *Journal of Agricultural Food and Chemistry,* 52, 4360-4367.

[15] Macheix, J.J., Fleuriet, A. and Billot, J. (1990). The main phenolics of fruits. In: Maxcheiz, J.J., Fleuriet, A. and Billot, J. (Eds.), *Fruit phenolics.* (378pp.). Boca Raton: CRC Press.

[16] Abe, L.T., Da Mota, R.V., Lajolo, F.M. and Genovese, M.I. (2007). Compostos fenólicos e capacidade antioxidante de cultivares de uvas Vitis labrusca L. e Vitis vinifera L. Ciência e Tecnologia de Alimentos, 27, 394-400.

[17] Fernandez-Panchon, M.S., Villano, D., Troncoso, A.M. and Garcia-Parrilla. (2008). Antioxidant activity of phenolic compounds: from in vitro results to in vivo evidence. *Critical Reviews in Food Science and Nutrition,* 48, 649-671.

[18] Stalikas, C.D. (2007). Extraction, separation, and detection methods for phenolic acids and flavonoids. *Journal of Separation Science*, 30, 3268-3295.

[19] Shi, J., Pohorly, J.E. and Kakuda, Y. (2003). Polyphenolics in grape seeds – biochesmistry and functionality. *Journal of Medicinal Food*, 6, 291-299.

[20] Wang, X., Tong, H., Chen, F. and Gangemi, J.D. (2010). Chemical characterization and antioxidant evaluation of muscadine grape. *Food Chemistry,* 123, 1156-1162.

[21] Amico, V., Napoli, E.M., Renda, A., Ruberto, G., Spatafora, C. and Tringali, C.(2004). Constituents of grape pomace from the Sicilian cultivar 'NerelloMascalese'. *Food Chemistry*, 88, 599-607.

[22] Spranger, I., Sun, B., Mateus, A.M., Freitas, V. and Ricardo-da-Silva, J.M. (2008). Chemical characterization and antioxidant activities of oligomeric and polymeric procyanidin fractions from grape seeds. *Food Chemistry,* 108, 519-532.

[23] Makris, D.P., Boskou, G. and Andrikopoulos, N.K. (2007). Polyphenolic content and in vitro antioxidant characteristics of wine industry and other agri-food solid waste extracts. *Journal of Food Composition and Analysis*, 20, 125-132.

[24] Yi, C., Shi, J., Kramer, J., Xue, S., Jiang, Y., Zhang, M., Ma, Y.and Pohorly, J. (2009). Fatty acid composition and phenolic antioxidants of winemaking pomace powder. *Food Chemistry*, 114, 570–576.

[25] González-Centeno, M.R., Rosselló, C., Simal, S., Garau, M.C., López, F. and Femenia, A. (2010). Physico-chemical properties of cell wall materials obtained from ten grape varieties and their byproducts: grape pomaces and stems. *LWT – Food Science and Techonology*, 43, 1580-1586.

[26] Rababah, T.M., Ereifej, K.I., Al-Mahasneh, M.A., Ismaeal, K., Hidar, A. and Yang, W. (2008). Total phenolics, antioxidant activities, and anthocyanins of different grape seed cultivars grown in Jordan. *International Journal of Food Properties*, 11, 472-479.

[27] Leaf, A. and Weber, P.C. (1988). Cardiovascular effects of n3 fatty acids. The New England Journal of Medicine, 318, 549–557.

[28] Banni, S. and Martin, J.C. (1998). Conjugated linoleic acid and metabolites. In J.L. Sebedio (Ed.), *Trans fatty acids in human nutrition* (pp.261–302). Dundee: The Oily Press.

[29] Fernandez, M.A., De Las Heras, B., Garcia, M.D., Saenz, M.T. and Villar, A. (2001). New insights into mechanism of action of the anti-inflammatory triterpine Iupeol. *Journal of Pharmacy and Pharmacology,* 53, 1533–1539.

[30] Hata, K., Hori, K. and Takahashi, S. (2002). Differentiation and apoptosis-inducing activities by pentacyclic triterpenes on a mouse melanoma cell line. *Journal of Natural Products*, 65, 645–648.

[31] Saleem, M., Alam, A., Arifin, S., Shah, M.S., Ahmed, B. and Sultana, S. (2001). Lupeol, atriterpene, inhibits early responses of tumor promotion induced by benzoyl peroxide in murine skin. *Pharmacological Research*, 43, 127–134.

[32] Llobera, A. and Cañellas, J. (2007). Dietary fibre content and antioxidant activity of Manto Negro red grape (Vitis vinifera): pomace and stem. *Food Chemistry*, 101, 659-666.

[33] Lafka, T.I., Sinanoglou, V. and Lazos, E.S. (2007). On the extraction and antioxidant activity of phenolic compounds from winery wastes. *Food Chemistry*, 104, 1206-1214.

[34] Rockenbach, I.I., Rodrigues, E., Gonzaga, L.V., Caliari, V., Genovese, M.I., Gonçalves, A.E.S.S. and Fett, R. (2011). Phenolic compounds content and antioxidant activity in pomace from selected red grapes (Vitis vinifera L. and Vitis labrusca L.) widely produced in Brazil. *Food Chemistry*, 127, 174-179.

[35] Thimothe, J., Bonsi, I.A., Padilla-Zakour, O.I. and Koo, H. (2007). Chemical characterization of red wine grape (Vitis vinifera and Vitis interspecific hybrids) and pomace phenolic extracts and their biological activity against Streptococcus mutans. *Journal of Agricultural and Food Chemistry*, 55, 10200–10207.

[36] Fuleki, T. and Silva, J.M.R. (1997). Catechin and procyanidin composition of seeds from grape cultivars grown in Ontario. *Journal of Agricultural and Food Chemistry*, 45, 1156-1160.

[37] Vinson, J.A., Dabbagh, Y.A., Serry, M.M. and Jang, J. (1995). Plant flavonoids, especially tea flavonols, are powerfull antioxidants using an in vitro oxidation model for heart disease. *Journal of Agricultural and Food Chemistry, 43, 2800-2802.*

[38] *Hogan, S., Cann*ing, C., Sun, S., Sun, X. and Zhou, K. (2010). Effects of grape pomace antioxidant extract on oxidative stress and inflammation in diet induced obese mice. *Journal of Agricultural and Food Chemistry*, 58, 11250-11256.

[39] Jeong, C.H., Kwak, J.H., Kim, J.H., Choi, G.N., Kim, D. and Heo, H.J. (2011). Neuronal cell protective and antioxidant effects of phenolics obtained from Zanthoxylum piperitum leaf using in vitro model system. *Food Chemistry,* 125, 417-422.

[40] Choudhury, D., Das, A., Bhattacharya, A. and Chakrabarti, G. (2010). Aqueous extract of ginger shows antiproliferative activity through disruption of microtubule network of cancer cells. *Food and Chemical Toxicology,* 48, 2872-2880.

[41] Lu, Y. and Foo, L.Y. (1999). The polyphenol constituents of grape pomace. *Food Chemistry,* 65, 1-8.

[42] Wang, H., Nair, M.G., Strasburg G.M., Chang, Yu-Chen, Booren, A.M., Gray,J.I. and Dewitt, D.L. (1999). Antioxidant and antiinflammatory activities of anthocyanins and their aglycon, cyanidin from tart cherries. *Journal of Natural Products*, 62, 294–296.

[43] Kähkönen, M.P. and Heinonen, M. (2003). Antioxidant activity of anthocyanins and their aglycons. *Journal of Agricultural and Food* Chemistry*, 51, 628–633.*

[44] Kähkönen, M.P., Heinämäki, J., Ollilainen, V. and Heinonen, M. (2003). Berry anthocyanins: isolation, identification and antioxidant activities. *Journal of the Science of Food and Agriculture*, 83, 1403–1411.

[45] Monrad, J.K., Howard, L.R., King, J.W., Srinivas, K. and Mauromoustakos, A. (2010). Subcritical solvent extraction of anthocyanins from dried red grape pomace. *Journal of Agricultural and Food Ckemistry*, 58, 2862-2868.

[46] Souquet, J.M., Labarbe, B., Guernevé, C.L., Cheynier, V.and Moutounet, M. (2000). Phenolic composition of grape stems. *Journal of Agricultural and Food Chemisty*, 48, 1076-1080.

[47] Gray, J. I., Gomaa, E. A. and Buckley, D.J. (1995). Oxidative quality and shelf life of meats. Meat Science, *Great Britain*, 34, S111-S123.

[48] Osawa, C.C., Felício, P.E.E. and Gonçalves, L.A.G. (2005). TBA test applied to meats and their products: Traditional, modified and alternative methods. *Química Nova,* 28, 655-663.

[49] Allen, J.C. and Hamilton, R.J. (1994). *Rancidity in foods*. 3rd ed. London: Blackie Academic.

[50] Halliwell, B. The characterization of antioxidants. (1995). *Food and Chemistry Toxicology*, 33, 601-617.

[51] Jayaprakasha, G.K., Singh, R.P. and Sakariah, K.K. (2001). Antioxidant activity of grape seed (Vitis vinifera) extracts peroxidation models in vitro. *Food Chemisty*, 73, 285-290.

[52] Dziezak, J.D. Antioxidants - the ultimate answer to oxidation. (1986). *Food Technology*,.40, 94-101.

[53] Labuza, T.P., Heidelba, N.D., Silver, M. and Karel, M. (1971). Oxidation at intermediate moisture contents. *Journal of the American Oil Chemists Society*, 48, 86-89.

[54] Frankel, E.N. (1998). *Lipid oxidation*. Dundee: The Oily Press.

[55] Schuler, P. Natural antioxidants exploited commercially. (1990). In: Hudson, B.J.F. (Ed.), *Food Antioxidants*. (317pp. 88-70). London: Elsevier Applied Science.
[56] Madhavi, D.L. (1995). Food antioxidant: sources and methods of evaluation. New York: Marcel Decker.
[57] Ahmad, J. I. Free radicals and health: Is vitamin E the answer? (1996). *Food Science and Technology*, 10, 147-152.
[58] Ahn, D. U., Olson, D.G., Lee, J.I., Jo, C., Wu, C. and Chen, X. (1998). Packaging and irradiation effects on lipid oxidation and volatiles in pork patties. *Journal of Food Science*, 63, 15-19.
[59] González-Paramás, A. M., Esteban-Ruano, S., Santos-Buelga, C., Pascual-Teresa, S. and Rivas-Gonzalo, J. C. (2004). Flavanol content and antioxidant activity in winery byproducts. *Journal of Agricultural and Food Chemistry*, 52, 234-238.
[60] Shirahigue, L. D., Plata-Oviedo, M., Alencar, S. M., Regitano-D'Arce, M. A. B., Vieira, T. M. F. S., Oldoni, T. L. C. and Contreras-Castillo, C. J. (2010). Wine industry residue as antioxidant in cooked chicken meat. *International Journal of Food Science and Technology*, 45, 863–870.
[61] Selani, M.M., Contreras-Castillo, C.J., Shirahigue, L.D., Gallo, C.R., Plata-Oviedo, M. and Montes-Villanueva. (2011). Wine industry residues extracts as natural antioxidants in raw and cooked chicken meat during frozen storage. *Meat Science*, 88, 397-403.
[62] Sasse, A., Colindres, P. and Brewer, M.S. (2009). Effect of natural and synthetic antioxidants on the oxidative stability of cooked frozen pork patties. *Journal of Food Science*, 74, S30-S35.
[63] Bañón, B, Díaz, P., Rodríguez, M., Garrido, M.D.and Price, A. (2007). Ascorbate, green tea and grape seed extracts increase the shelf life of low sulphite beef patties. *Meat Science*, 77, 626-633.
[64] Ogawa, M. and Maia, E. L. (1999). *Manual de pesca*. São Paulo: Livraria Varela.
[65] Alonso, A. M., Guillén, D.A, Barroso, C.G., Puertas, B. and Garcia, A. (2002). Determination of antioxidant activity of wine by products and its correlation with polyphenolic content. *Journal of Agriculture and Food Chemistry,* 50, 5832-5836,
[66] Torres, J.L., Varela, B., García, M.T., Carilla, J., Matito, C., Centelles, J.J., Cascante, M., Sort, X. and Bobet, R. (2002). Valorization of grape (Vitis vinifera) byproducts. Antioxidant and biological properties of polyphenolic fractions differing in procyanidin composition and flavonol content. *Journal of Agricultural and Food Chemistry*, 50, 7548-7555.

[67] Pazos, M., Gonzáles, M.J., Gallardo, J.M. and Torres, J.L. (2005). Preservation of the endogenous antioxidant system of fish muscle by grape polyphenols during frozen storage. *European Food Research and Technology*, 220, 514-519.

[68] Pazos, M., Gallardo, J.M., Torres, J.L. and Medina, I. (2005). Activity of grape polyphenol as inhibitors of the oxidation of fish lipids and frozen fish muscle. *Food Chemistry*, 92, 547-557.

[69] Yagci, S. and Gogus, F. 2009. Effect of incorporation of various food by-products on some nutritional properties of Rice-based extruded foods. *Food Science and Technology International*, 15, 571-581.

[70] Peng, X., Ma, J., Cheng, K-W., Jiang, Y., Chen, f. and Wang, m. (2010). The effects of grape seed extract fortification on the antioxidant activity and quality attributes of bread. *Food Chemistry*, 119, 49-53.

[71] Bussolo, T.M. and Thomé, V.A. (2008). Efeito dos extratos de bagaços de uvas Niágara e Isabel na reologia das massas e na qualidade do pão de forma. Trabalho de Conclusão de Curso (Graduação). Curso Superior de Tecnologia em Processamento de Alimentos Vegetais. Universidade Tecnológica Federal do Paraná, Campo Mourão, Paraná, Brazil.

[72] Shan, B., Cai, Y., Brooks, J.D. and Corke, H. (2011). Potential application of spice and herbs extracts as natural preservatives in cheese. *Journal of Medicinal Food*, 3, 284-290.

[73] Hwang, J.Y., Shyu, Y.S. and Hsu, C.K. (2009). Grape wine lees improves the rheological and adds antioxidants properties to ice cream. *LWT – Food Science and Technology*, 42, 312-318.

In: Grapes ISBN 978-1-61470-950-3
Editors: R. P. Murphy et al., pp. 79-106

Chapter 3

EVALUATION OF GRAPE SEEDS AS A SOURCE OF ADDED VALUE NATURAL ANTIOXIDANTS: AQUEOUS EXTRACTION OF HIGH MOLECULAR WEIGHT PHENOLICS

Veronica Sanda Chedea*[*1,2], Sonia Moussouni[3], Carmen Socaciu[4] and Panagiotis Kefalas[3]

[1]Laboratory of Animal Biology, National Research Development Institute for Animal Biology and Nutrition (IBNA), Balotesti, Ilfov, 077015, Romania

[2]Department of Life Science and Biotechnology, Shimane University, 1060 Nishikawatsu, Matsue, Shimane 690-8504, Japan

[3]Department of Food Quality and Chemistry of Natural Products, Mediterranean Agronomic Institute of Chania, C entre International de Hautes Etudes Agronomiques Méditerraneénnes, Chania, 73100 Chania, Crete, Greece

[4]Department of Chemistry and Biochemistry, University of Agricultural Sciences and Veterinary Medicine, 400372 Cluj-Napoca, Romania

---

[*] E-mail:chedeaveronica@hotmail.com

ABSTRACT

Water extracts of a mixture of seeds originating from red and white Hellenic native and international *Vitis vinifera* varieties cultivated in Greece were screened for their characteristic polyphenols and antioxidant activity in polar and bulk oil media. By LC-DAD-MS in ESI+ mode procyanidine dimers, trimers, tetramers, pentamers and hexamers were tentatively identified. A procyanidin trigalloylated nanomer and a trigalloylated octamer were identified when the sedimentation residues from the first extraction were reextracted in ethyl acetate. The aqueous extracts contain more total polyphenols than the ethyl acetate ones but generally the ethyl acetate extracts have a better iron reducing power, antiradical activity and prevent better the oxidation of sunflower oil and tocopherol free sunflower oil. The tocopherols from the unpurified oil seem to have an antagonistic action against the polyphenolic extracts, reducing the antioxidant activity of those.

Keywords: grape seed extract, procyanidin oligomers, antioxidant activity.

LIST OF ABBREVIATIONS

grape seed extract, GSE;
liquid chromatography coupled with mass spectrometry in tandem LC-MS/MS;
2,2-diphenyl-β-picrylhydrazyl, DPPH;
ethylenediamine tetraacetic acid, EDTA;
Ferric Reducing Antioxidant Power, FRAP;
liquid chromatography diode array detection coupled with mass spectroscopy, LC-DAD-MS;
aqueous red grape seed extract, RW;
aqueous white grape seed extract WW;
ethyl acetate extract of RW sedimentation residues taken up in methanol, REM;
ethyl acetate extract of WW sedimentation residues taken up in methanol, WEM;
ethyl acetate extract of WW sedimentation residues taken undissolved in methanol and so taken up in water, WEW;
total polyphenols, TP;
gallic acid equivalents, GAE;

antiradical activity, A_{AR};
Trolox® equivalents TRE;
chemiluminescence intensity at the absence of antioxidant, I_o;
concentration of compound, which is required to decrease I_o intensity by 50%, IC_{50};
retention times, t_R;
(epi)catechin, (E)C;
(epi)catechin gallate, (E)CG;
electrospray ionisation, ESI;
gallic acid, GA;
ethyl acetate, EtOAc;

1. INTRODUCTION

Grape seeds are waste products of the winery and grape juice industry. These seeds contain lipid, protein, carbohydrates, and 5-8% polyphenols, depending on the variety. Grape seed polyphenols contain flavan-3-ols as monomers (catechin, epicatechin, gallocatechin, epigallocatechin and epicatechin 3-O-gallate) but also as procyanidin dimers, trimers, and highly polymerized procyanidins, aside from phenolic acid precursors (gallic acid). For this reason, the grape seed extract (GSE) is to be considered as a powerful antioxidant that prevents premature aging and disease (Schewe et al. 2001; Schewe et al. 2002).

The most abundant phenolic compounds isolated from grape seeds are catechin, epicatechin, and procyanidins (Shi et al. 2003). Catechin is usually the most important individual flavanol in both grape skins and seeds, although epicatechin is also usually well represented. Some grape varieties display similar levels of both monomers, or even higher proportion of epicatechin (Gonzales-Manzano et al. 2004; Chedea et al. 2011).

Procyanidin B_1 has been reported to be the main oligomer in skins (Mateus et al. 2001), whereas all C_4–C_8 procyanidin dimers (i.e., B_1–B_4) are usually found in seeds, the procyanidin B_2 is normally the most abundant one (Jordão et al. 2001). Grape seed proanthocyanidins constitute a complex mixture consisting of procyanidins and procyanidin gallates (Hayasaka et al. 2003). Levels of galloylated flavan-3-ols are more important in seeds than in skins (Jordão et al. 2001).

The proanthocyanidins from GSE, have shown promising chemopreventive and/or anticancer properties in various cell culture and animal models (Surh 2003). Also it was found that GSE prevents azathioprine toxicity in rats (El-Ashmawy et al. 2010). Azathioprine is an important drug commonly used in the therapy of autoimmune system disorders.

The iron-chelating activity of GSE minimizes its pro-oxidant activity and delays 6-hydroxydopamine auto-oxidation to provide cytoprotection in PC-12 cells (Wu et al. 2010). GSE enhanced the antioxidant status and decreased the incidence of free radical-induced lipid peroxidation in the central nervous system of aged rats (Balu et al. 2005). Using LC-MS/MS, (+)-catechin and (−)-epicatechin were identified in the brain conclusively, suggesting that GSE catechins cross the blood brain barrier and may be responsible for the neuroprotective effects of GSE (Prasain et al. 2009).

The proanthocyanidins from GSE, have been reported to strengthen collagen-based tissues by increasing collagen cross-links. GSE positively affects the demineralization and/or remineralization processes of artificial root caries lesions, most likely through a different mechanism than that of fluoride and so, GSE may be a promising natural agent for non-invasive root caries therapy (Xie et al. 2008).

It is estimated that around 13% of the total weight of grapes used for the wine making results in grape pomace that is a by-product in this process (Torres et al. 2002). Dealing with the problem of this waste disposal the wine producers have to balance two major issues given by this high polyphenolic composition: the properties of germination's inhibition by this compounds which may have an adverse environmental impact (Morthup et al. 1998) and the beneficial effects on human health of polyphenols (Shi et al. 2003; Torres et al. 2002).

If conveniently processed, the grape pomace could provide useful products that may balance out waste treatment costs (Amico et al. 2004). One way of using the potential of the grape pomace is the isolation of seeds and extraction of the polyphenols since the total extractable phenolics in grape are present in about maximum 10% in pulp, 60-70% in the seeds and 28-35 % in the skin (Shi et al. 2003).

The goal of the current work is to investigate *a simple and practical approach for the valorization* of the seeds by using water and recyclable ethyl acetate as extraction solvents. In this view their antioxidant activity was determined by the Ferric Reducing Antioxidant Power, as well as by the DPPH radical and Luminol chemiluminescence methods. Behaviour in bulk oil medium was evaluated by the Rancimat test.

The total polyphenolic content was measured by the Folin–Ciocalteu method. The polyphenolic composition of the five GSE samples was investigated by LC-DAD-MS and correlation between structures and activity was sought.

2. MATERIALS AND METHODS

2.1. Chemicals

2,2-diphenyl-β-picrylhydrazyl (DPPH·) radical was from Sigma Aldrich Steinheim Germany. Boric acid was purchased from Applichem (Germany), cobalt (II) chloride hexahydrate, hydrogen peroxide 35% (v/v), acetic acid and phosphoric acid, Folin-Ciocalteu reagent, gallic acid, sodium carbonate and active carbon from Merck (Germany), luminol (3-aminophtalylhydrazide), ethylenediamine tetraacetic acid (EDTA), Trolox, silica gel and diatomic earth from Sigma (USA), methanol and hexane pro analysis from Fluka, (Germany).

Sunflower refined oil and powder table sugar are the ones commercially available in any food store.

2.2. Plant Material

All seed samples studied were from varieties selected to cover major parts of the Hellenic vineyard in order to have a specimen as representative as possible of the totality of grape varieties (*V. vinifera* sp.) cultivated in Greece. Analytical information about the origin and vineyard location is given in Table 1. The grapes used were harvested at optimum technological maturity.

The seeds from white varieties were mixed by 2 g from each one, thus obtaining the starting material for the white grape seed extracts; the same procedure was applied to the red varieties. Grape berries were manually deseeded and the seeds were frozen in liquid nitrogen immediately afterwards, and stored in the freezer (-20°C) until analysed.

2.3. Polyphenol Extraction

The seeds were ground with liquid nitrogen, and extracted for 1 h by adding boiled water (5g ground seeds/40ml water) on the seed powder. After sedimentation the supernatant was centrifuged for 10 min at 3000 rpm, filtered, aliquoted and kept at -20^0 C till further analysed, thus obtaining the extracts RW for red and WW for white varieties respectively.

The sedimentation residues were reextracted with ethyl acetate for 2 hours at room temperature. After 2 hrs the solvent was evaporated to dryness and the residue taken up in 4ml of methanol for the red grape seed sediment (REM extract) and 4ml of methanol for the white grape seed sediment (WEM extract) respectively. The undissolved residuals of the white grape seed sediment was taken in 2 ml H_2O (WEW extract).

Table 1. Origin and location within Greece of the white and red grape varieties of which the seeds were examined

Variety	Origin	Location[a]
	White grape varieties	
Asyrtiko	Santorini	Aegean Isles (S)
Athiri	Rodos	Aegean Isles (S)
Chardonnay	Kavala	Macedonia (N)
Greco Bianco	Attica	Sterea Ellada(C)
Kontokladi	Attica	Sterea Ellada (C)
Malagouzia	Chalkidiki	Macedonia (N)
Moschardinia	Zakinthos	Ionian Isles (S)
Mygdali	Attica	Sterea Ellada(C)
Roditis	Patra	Peloponnese (S)
Savvatiano	Attica	Sterea Ellada(C)
Skiadopoulo	Attica	Sterea Ellada(C)
Thiako	Attica	Sterea Ellada(C)
	Red grape varieties	
Pardala	Attica	Sterea Ellada (C)
Refosko	Ilia	Peloponnese (S)
Cabernet Sauvignon	Volos	Thessaly (C)
Grenache Rouge	Attica	Sterea Ellada (C)
Merlot	Chalkidiki	Macedonia (N)

Variety	Origin	Location[a]
Moschofilero	Mantinia	Peloponnese (S)
Aidani Mavro	Attica	Sterea Ellada (C)
Krasato	Rapsani	Thessaly (C)
Thrapsa	Messinia	Peloponnese (S)
Papadiko	Attica	Sterea Ellada (C)
Limniona	Karditsa	Thessaly (C)
Mavro Messenikola	Karditsa	Thessaly (C)
Mandilaria	Rhodes	Aegean Isles (S)
Agioritiko	Nemea	Peloponnese (S)
Vapsa	Attica	Sterea Ellada (C)
Karlachanas	Attica	Sterea Ellada (C)
Negoska	Goumenissa	Macedonia (N)
Xinomavro	Naoussa	Macedonia (N)
Araklinos	Attica	Sterea Ellada (C)
Muscat of Hamburg	Tirnavos	Thessaly (C)
Fileri	Messinia	Peloponnese (S)
Mavrodafni	Patra	Peloponnese (S)
Avgoustiatis	Zakinthos	Ionian Isles (S)
Sangiovese	Attica	Sterea Ellada (C)
Limnio	Chalkidiki	Macedonia (N)

[a] N, C and S denote northern, central and southern Greece, respectively.

2.4. Total Polyphenol Determination

Total polyphenols (TP) were determined using the Folin–Ciocalteau reagent, with the microscale protocol previously developed (Arnous et al. 2001, 2002). Gallic acid was employed as a calibration standard and results were expressed as gallic acid equivalents (mg GAE/ g of seeds). Before starting the TP determinations all extracts were centrifuged for 3 min at 10.000 rpm and the supernatant was used for the analysis.

In a 1.5 ml Eppendorf tube, 0.79 ml distilled water, 0.01 ml sample appropriately diluted (1:4 for RW and WW, 1:16 for REM and WEW, no dilution for WEM), and 0.05 ml Folin-Ciocalteu reagent were added and vortexed. After 1 min, 0.15ml of sodium carbonate (20%) were added, the mixture was vortexed and allowed to stand at room temperature in obscurity, for 120 min. The absorbance was read at 750 nm and the total polyphenol concentration was calculated from a calibration curve (r^2=0.9990), using gallic acid as standard (50-800 mg/l).

2.5. LC-DAD-MS Analysis of Grape Seed Extracts

For the LC-DAD-MS analysis 500 μl of the stock RW, WW, REM, WEM and WEW solutions were diluted in 1 ml methanol and filtered through SPARTAN 13, 0.45 μm (Whatman) filter discs prior to chromatographic separation.

The analysis was performed in the electrospray positive ion mode (ESI+) on an AQA mass spectrometer (Thermoquest/ Finnigan), coupled to a P4000LC pump (Finnigan) and a UV6000LP diode array detector (Finnigan). The separation was performed on a Synergi 4μ Hydro-RP 80A (150x2 mm) column by Phenomenex at a flow rate of 0.3 mL/min, the column being kept at 40ºC. The detection was monitored at 278 nm. The probe temperature was 350ºC, the probe voltage 4.00 kV and the collision induced fragmentation energy set at 20eV (CID). The mass range was set at 130-1546 amu and the scan rate was 0.8 scans/sec. The mobile phase consisted of A, acetic acid 2.5%, and B methanol. The elution was isocratic for 2 min at 100% A, then reached 100% B after 50 min and kept isocratic (100% B) for another 8 min.

2.6. Measurement of the Ferric Reducing Antioxidant Power (FRAP)

For the determination of the reducing power a protocol based on the ferric reducing antioxidant power (FRAP) assay was applied, as described previously (Benzie and Strain, 1996).

A fresh working FRAP solution was prepared by mixing 30mL of acetate buffer (300 mM, ph 3.6) solution, 3mL of TPTZ (2, 4, 6-tripyridyl-s-triazine) solution (10 mM), and 3mL of $FeCl_3$- $6H_2O$ solution (20 mM). GSE extracts (100μL) were allowed to react with 3000μL of the FRAP solution and absorbance was measured at 593nm (A_{593}) of the coloured product [ferrous tripyridyltriazine complex] at 0 min after vortexing. Thereafter, the samples were placed at 37°C in a water bath and absorption was measured again after 4 min.

For comparison reasons ascorbic acid was used as the calibration standard, because it is a well characterized natural reducing agent. 100μL of ascorbic acid solution (1 mM) were processed in the same way. Results were expressed as FRAP values using the following equation:

FRAP = (change in absorbance of extract from 0 to 4 min/ change in absorbance of standard from 0 to 4 min) × FRAP value of ascorbic acid

FRAP value of ascorbic acid has a stoichiometric value of 2, i.e. 1 mol ascorbic acid reduces 2 mol Fe^{3+} to Fe^{2+}.

2.7. Measurement of the Antiradical Activity (A_{AR})(DPPH Assay)

A DPPH• (2,2-diphenyl-β-picrylhydrazyl) radical-scavenging assay was employed as described by Arnous et al (2002) to determine the hydrogen donating ability of the redissolved extract. A volume of 975 μl DPPH• solution (60 μM in MeOH) was used. The reaction was started by the addition of 25 μl of extract followed by vortexing. The bleaching of DPPH• was followed at 515 nm at 25° C by reading the absorption after 30 min. The inhibition percentage (IP) of the DPPH· radical was calculated as follows:

$$IP = \frac{absorbance_{t=0min} - absorbance_{t=30min}}{absorbance_{t=0min}} \times 100$$

Results were expressed as Trolox® equivalents (mM TRE) using the following equation:

A_{AR} (mM TRE) = (IP – 1.803)/47.082

as determined from linear regression (y=47.082x+1.8031, r^2 = 0.9999), after plotting IP against known solutions of Trolox® concentration (0.08-1.28 mM).

All the above measurements were done for 5 concentrations (0.3mg GAE, 0.15 mg GAE, 0.075 mg GAE, 0.0375 mg GAE and 0.01875mg GAE) for each extract, expressing the results as EC_{50} (defined as the extract concentration which achieves a decrease of DPPH• absorbance to 50% of the initial value).

2.8. Chemiluminescence Measurements

A chemiluminescence method was used as described by Parejo et al. (2000). One milliliter of borate buffer solution (50 mM, pH 9) containing $CoCl_2.6H_2O$ (8.40 mg mL^{-1}) and EDTA (2.63 mg mL^{-1}) was first mixed with 0.1 mL of luminol solution (0.56 mM in borate buffer 50 mM, pH 9.00) and vortexed for 15 s. All the samples were dissolved in water, in concentrations ranging from 0.30 to 0.019 mg GAE.

An aliquot of 0.025 mL of sample and 0.025 mL of H_2O_2 aqueous solution (5.4 mM) were then added into the test tube, the mixture was rapidly transferred into a glass cuvette, and the chemiluminescence intensity (I_o) was recorded.

The instantaneous reduction in I_o elicited by the addition of the sample was recorded as I and the ratio (I_o/I) was calculated and plotted vs. concentration (mg GAE) of the sample. The concentration of sample (IC_{50}), which is required to decrease I_o intensity by 50%, was also calculated. For all measurements, a fluorimeter (model 6200, Jenway Ltd., Gransmore, Essex, UK) was used, keeping the lamp off and using only the photomultiplier of the apparatus.

2.9. Rancimat Measurements (Conductivity)

The conductivity measurements were done on a Methrom 679 Rancimat, Herisau Switzerland apparatus. 400 ppm extract were added in 3 g unpurified and respectively purified sunflower oil, the conductivity was measured for 20 h having the samples heated at 100° C with an air flow of 20 L/h.

2.9.1. Elution of Sunflower Oil on a Preparative Chromatographic Column to Remove Vitamin E

The column (60cm x 2cm) was filled up with : 10g activated silica gel+ diatomic earth (1:1, w/w) (at the bottom), 5g diatomic earth + active carbon (1:2, w/w), 20 g diatomic earth + powder table sugar (1:2, w/w), 10g activated silica gel+diatomic earth (1:1, w/w) (top of the column). The column was conditioned with hexane. Then 50 g of sunflower oil were dissolved in 250 ml hexane and eluted through the column. Vit E is retained onto the column. The resulting hexane solution was evaporated by using a rotary evaporator and the resulting oil was kept at -18°C.

2.10. Determination and Statistics

All measurements were made at least in triplicate in order to calculate standard deviations, and correlations were established using linear regression analysis.

3. Results and Discussion

The same seed samples from 12 white and 25 red international and Hellenic native grape varieties (*Vitis vinifera*), that we have taken for analysis were individually screened for their polyphenolic composition by Guendez et al. (2005). The authors have determined mainly polyphenols of low molecular weight, including gallic acid, catechin, epicatechin, epicatechin gallate, epigallocatechin, epigallocatechin gallate and the procyanidins B_1 and B_2. Average values of total content for white and red varieties (376 and 388 mg/100 g seeds, respectively) were very similar (Guendez et al. 2005). Comparable results were observed with respect to the individual polyphenol content with seeds from red varieties being, in general, slightly richer. The predominant flavanol monomer in white and red varieties was catechin (which accounted for 50.5 and 49.3%, respectively, of the total content), whilst gallic acid and epigallocatechin were the constituents showing the lowest content, respectively (Guendez et al. 2005). Our study went further into the estimation of antioxidant activity of water and ethyl acetate extracts rich in procyanidins that have a higher degree of polymerization and galloylation.

3.1. Polyphenol Composition of RW and WW Extracts Determined by LC-DAD-MS

Figure 1 presents the HPLC chromatograms of RW (A) and WW (B) aqueous extracts.

Table 2 presents the identified compounds from the aqueous extracts RW and WW, associated with their peak retention times (t_R) and *m/z* values.

Table 2. Identification of phenolic species contained in aqueous extracts of grape seeds, RW and WW

Peak	Retention time (min) RW *WW*	λ_{max} (nm)	m/z	Other positive ion fragments (m/z) $[M+H]^+$, $[M+Na]^+$, $[M+H_2O]^+$	Tentative fragment identification
1	3.48; *3.63*	270			Gallic acid
2	7.04; *7.67*	240;298 *254;296*	339		Hydroxybenzoic acid derivatives (dimer of 2,3,4-trihydroxybenzoic acid)
3	10.48; *11.18*	240;278	579	*601=578+23*	Procyanidine dimer
4	11.63; *12.16*	240;278	291		Catechin
5	11.80; *12.28*	240;278	291		Epicatechin
6	13.94; *14.28*	242;278	579		Procyanidine dimer
7	14.95	240;278	867		Procyanidine trimer
8	*14.99*	*240;278*	*579*		*Procyanidine dimer*
9	15.92 *16.19*	240;278	867	291, 579	Procyanidine trimer
Zone 1	16.5-20.00 *16.5-20.00*	242; 278	1443 1307	1155, 1307, 1155, 1019, 867 1460=1442+18	Pentamer (epi)cat Gallate Tetramer 3 (epi)cat+1 ECG
Zone 2	20.00-23.00	240;278	1171 1307 1019	867=3(E)C, 1171=1(E)C+2 ECG 1019= 3(E)C+1 ECG 1019=2(E)C+1 ECG	Digallate Hexamer 4(E)C+2 ECG
10	23.39	240;278	883	731=441(ECG) +289 (EC)+1, 515=442+23+18 +32	Dimer ECG

Italic characters correspond to the WW chromatogram.

The main grape seed polyphenols, the flavan-3-ols and procyanidins, show no individual specific UV-Vis absorbtion maxima and all have a maximum at 280 nm, so the UV-Vis spectra can not help distinguish the catechin type compounds either as monomers, dimers, trimers or higher oligomers.

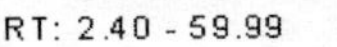

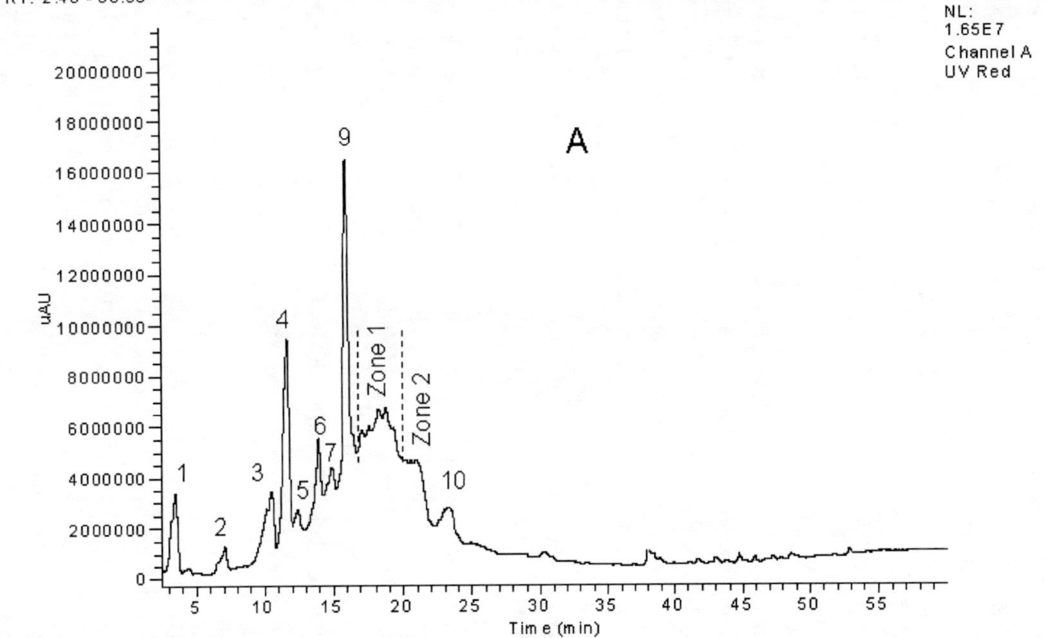

Figure 1. HPLC chromatograms for RW (A) and WW (B) extracts obtained at 278 nm. The peaks were tentatively assigned as follows: 1-Gallic acid, 2-Hydroxybenzoic acid derivatives, 3- Procyanidine Dimer, 4-Catechin, 5-Epicatechin, 6- Procyanidine Dimer, 7- Trimer, 8- Procyanidine Dimer, 9-Trimer EC, Zone 1- Pentamer of EC and MonogallateTetramer (3EC+1ECG), Zone 2- Digallate Hexamer (4 EC+2 ECG), 10-Digallate Dimer.

Nevertheless, we obtained MS evidence at least for the presence, of catechin and epicatechin, dimeric, trimeric, tetrameric and pentameric proanthocyanidins.

The attribution of the *m/z* value was given in concordance with literature data and generally $[M+H]^+$=291 *m/z* indicates the monomeric structures - (epi)catechin (E)C, $[M+H]^+$ =443 *m/z* the monomeric form of (epi)catechin gallate (E)CG, $[M+H]^+$=579 *m/z* an (E)C dimer, $[M+H]^+$=867 *m/z* a trimeric proanthocyanidin, $[M+H]^+$=1155 *m/z* a tetrameric structure $[M+H]^+$=1307 *m/z* a tetrameric structure of 3 EC units and 1 ECG, and $[M+H]^+$=1443 *m/z* a pentameric one (Pati et al. 2006; Wu et al. 2003; Amico et al. 2004, Núñez et al 2006).

The analysis of procyanidins in grape seeds is complex and has relied on reverse-phase HPLC with UV detection at 280 nm (Oszmianski and Sapis 1989, Santos-Buelga et al. 1995; Fuleki and Ricardo da Silva 1997; Sun et al. 1999). This method is successful in separating oligomers of equivalent molecular mass into their isomers up to tetramers. The separation of the higher oligomers (>tetramer) is difficult because the number of isomers concomitantly increases with the degree of polymerization, producing a very broad and unresolved UV-absorbing peak late in the chromatogram (Núñez et al. 2006).

Gallic acid was identified against external standard based on its t_R (t_R= 3.48 min for RW and t_R=3.63 min for WW) and UV-Vis spectrum maxima, λ_{max}=270. At min t_R= 7.04 for RW and t_R=7.67 for WW, with λ_{max}=240, 298 (RW) respectively λ_{max}=254, 296 (WW) the peak was identified as a hydroxybenzoic derivative. Rubilar et al. (2007) identified also in an ethanolic extract of almond hulls hydroxybenzoic derivatives eluting after gallic acid, having λ_{max}=258, 294 nm and m/z=339.

Peak 3 (t_R= 10.48 min for RW and t_R=11.18 min for WW) has $[M+H]^+$ = 579 *m/z* corresponding to a procyanidine dimer and is tentatively assigned as being procyanidine B_1 in accordance with the literature data (Guendez et al. 2005). Peak 4 (t_R= 11.63 min for RW and t_R=12.16 min for WW) and peak 5 (t_R= 11.80 min for RW and t_R=12.28 min for WW) have the same UV-Vis and MS spectra of catechin and its diastereoisomer, epicatechin. The literature data (Shi et al. 2003; Gomez-Alonso et al. 2007), present the elution order for the two isomers as catechin eluting first followed by epicatechin, we also assigned peak 4 as catechin and peak 5 as epicatechin.

For the chromatogram of the RW extract from approximately 16.5 min to 20 min a very broad and unresolved UV-absorbing peak named Zone 1 (Table 2) was recorded. A narrower unresolved UV-absorbing peak was identified as

Zone 2 for the same extract, from 20 min to 23 min. In case of WW only Zone 1 was observed. Based on MS fragmentation Zone 1, for both RW and WW, indicates the presence of pentamers and tetramers. The pentamer identified is formed by polymerization of 5 molecules of EC having $[M+H]^+ = 1443$ *m/z* and the structure with $[M+H]^+ = 1307$ *m/z* corresponds to a tetramer monogalloylated of 3 molecules of EC and one of ECG (Núñez et al. 2006). Comparing Zone 1 of RW with that of WW it was observed that in case of WW the proportion of EC pentamer is higher than for RW, a rough estimation can indicate for RW 60% pentamer and 40% tetramer and for WW 90% pentamer and 10% tetramer.

Zone 2 of RW was tentatively attributed to procyanidine hexamers and mainly a hexamer digalloylated formed from 4 EC molecules and 2 of ECG. For higher oligomers, the molecular ions could not be detected due to the limited scanning range (100-1500 Da) of the mass spectrometer used. However, the identification of the hexamer was done based on the most abundant ions of the spectrum for fraction Zone 2 and which were at m/z 867 indicating a trimer of EC, m/z 1171 of a digallate trimer of 1 EC and 2 ECG and m/z 1019 of 2 EC and 1ECG gallate trimer (Núñez et al. 2006). In the case of Zone two the degree of galloylation is higher than for Zone 1. In all grape varieties and vintages studied by Núñez et al. (2006), as the degree of polymerisation increased the levels of non-galloylated flavan-3-ols decreased. For t_R= 23.69 min for RW the compound was identified as ECG dimer ~~of~~ having $[M+H]^+$= m/z 883 and fragmentation ions of m/z 731 and m/z 515. This peak was not present in the chromatogram of WW extract.

3.2. Polyphenol Composition of REM, WEM and WEW Extracts Determined by LC-DAD-MS

Figure 2 presents the HPLC chromatograms of REM, WEM and WEW ethyl acetate extracts. Also for the three extracts UV-Vis spectra are shown indicating a difference in structure for the peaks eluting at approximately t_R=38.00 min.

If for WEM only one main peak was separated with t_R=38.03 min, for REM and WEW three compounds eluted at t_R=38.09 min, t_R=38.54 min and t_R=38.90 min, and t_R=37.90 min, t_R=38.28 min and t_R=38.80 min respectively (Figure 2). Based on the UV-Vis spectra and MS ionic fragmentation these peaks were attributed to galloylated procyanidins. As the three peaks have the same base, we concluded that isomeric compounds having different structural disposition of their flavan-3-ol units were separated.

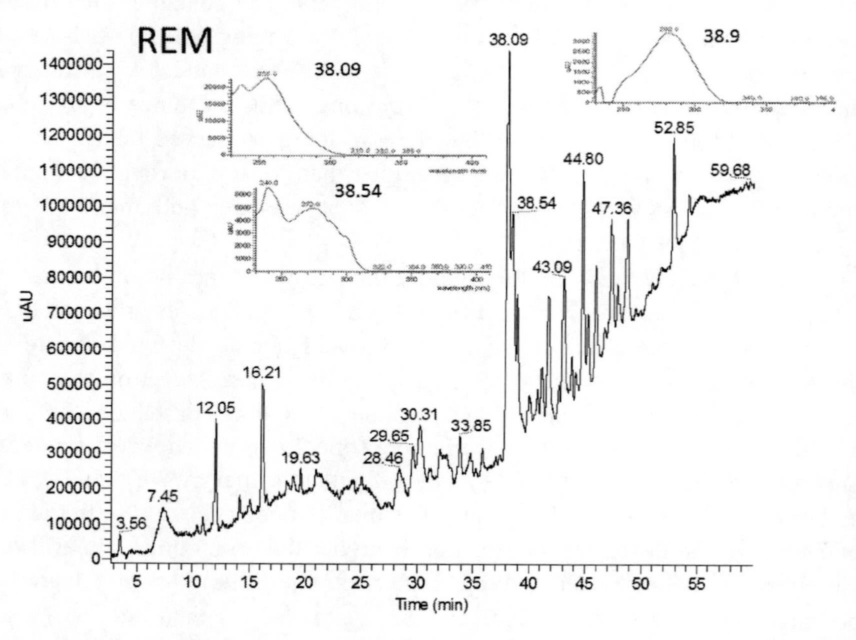

Figure 2. HPLC chromatograms for REM, WEM and WEW extracts obtained at λ=278 nm. Also for the three extracts UV-Vis spectra are shown for the peaks eluting at approximately at t_R=38.00 min. If for WEM only one main peak was separated with t_R=38.03 min, for REM and WEW three compounds eluted at t_R=38.09 min, t_R=38.54 min and t_R=38.90 min, and t_R=37.90 min, t_R=38.28 min and t_R=38.80 min respectively. The peaks eluting at above t_R were tentatively assigned as: procyanidin trigalloylated nanomer isomeric structures for REM and WEM extracts and a trigalloylated octamer isomers for WEW extract.

The UV spectra presented in Figure 2 support this idea. For the RW and WW extracts the catechin, epicatechin and procyanidinic oligomers, the UV spectra presented common maxima at about 240 nm and 278 nm so also in the case of REM,WEM, WEW extracts these UV maxima would indicate a procyanidinic condensation of the constitutive units.

The major peak is the one at t_R=38.09 min for REM, 38.03 min for WEM and 38.28 min for WEW. For WEM the most abundant peak t_R=38.03 min shows a fragment at m/z 1528. In their analysis of procyanidin polymers from grape seeds, Hayasaka et al. (2003), have identified the nanomer procyanidin trigallate with $[M-2H]^{2-}$=1525 and so we tentatively attributed the peak from

t_R=38.03 in the case of WEM extract to this compound. A difference of 5 may be -18 (a H_2O molecule) +23(Na^+) in ESI+. In ESI- it is possible to have -18 + 23 - 1 -1=3 (Tong et al. 1999) then 1528-1525=3.

M/z 1537 was registered for peaks eluting at t_R=38.09 min and t_R=38.54 min in the case of REM chromatogram. An ionization mechanism as presented bellow will determine us to tentatively identify these compounds as isomers of the same compound like in the case of WEM extract t_R=38.03 min based on the fragment with $[M-2H]^{2-}$=1525 of the procyanidin nanomer trigallate as determined by Hayasaka et al. (2003). The liberated formaldehyde has a MW of 30 (Figure 3). Then, a difference of 12 is 18 – 30 (+H_2O-HCHO); formation of a water adduct and elimination of HCHO, 1537 – 1525 = 12.

In the case of WEW extract, m/z of 1399, 1433 and 1409 for the 3 peaks eluting at t_R=37.9 min, t_R=38.28 min and respectively t_R=38.80 min were registered. We attributed these compounds to fragments of three isomeric structures of a procyanidine octamer trigallate. Hayasaka et al. (2003) have identified the octamer procyanidin trigallate with $[M-2H]^{2-}$=1380.6. In our case for m/z=1399 the elimination of one water molecule would leads to m/z=1381 and an elimination of CO from m/z=1409 leads to the same m/z=1381. An assumption of simultaneous elimination of two HCHO and one CO followed by adduct formation with two molecules of water leads to a difference of 52 amu which relates m/z 1433 to m/z 1381 (Hayasaka et al. 2003), i.e. 1433-52=1381.

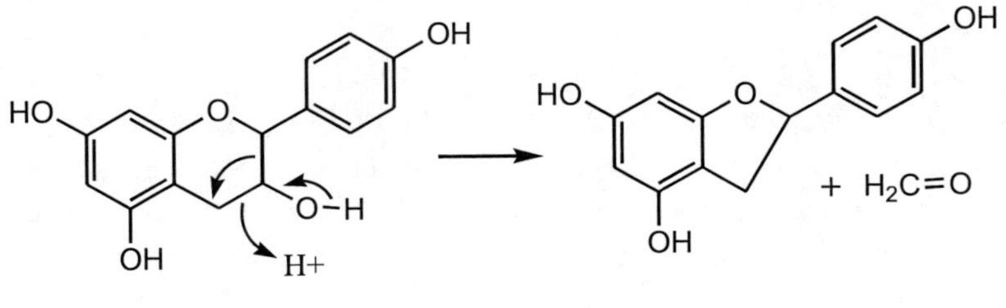

Figure 3. The presumed HCHO elimination in ESI+ mode.

The presence of gallocatechins and gallocatechin gallates that are hidden under other larger peaks became evident. The composition of the grape seed procyanidins is very complex. As the molecular weight increases, the number of isomers becomes so large that the separation and detection of individual isomers become almost impossible. Numerous diastereomers consisting of catechin, epicatechin, and their corresponding galloylated derivatives exist at each degree of polymerization (Yang and Chien, 2000). Hayasaka et al. (2003)

have also analysed in grape seeds from *Shiraz* variety berries, procyanidins with different degrees of polymerization (up to 28) and galloylation (up to 8).

As the degree of polymerisation of eluted procyanidins increased, the molecular ion peak signals becomes smaller and approaches the background noise - the ionization efficiency decreasing with the increasing of molecular size- makes difficult their identification (Freitas et al. 1998, Nunez et al. 2006). An apparent lower response could also result from a decreased ionization efficiency of the galloylated forms as compared to the non-galloylated forms (Nunez et al. 2006).

3.3. Total Polyphenols Content and Antioxidant Properties of RW, WW, REM, WEM and WEW

The TP content of the studied extracts as well as their antioxidant activity in aqueous environment, FRAP, A_{AR} assessed by DPPH• test and chemiluminescence measurements, and in lipidic medium determined by the Rancimat method for both unpurified and purified sunflower oil, are presented in Table 3.

3.3.1. Ferric Reducing Antioxidant Power (FRAP)

In this study the FRAP of GSE extracts and GA in decreasing order is: *WEW > REM>RW>WW>WEM>GA* (see Table 3).

A very high FRAP was measured for WEW extract (151.58±5.51). A 5.5 times lower FRAP was registered for REM (26.90±2.05) followed by RW (21.36±3.32) and WW (18.28±0.66), the difference between the last two extracts being statistically nonsignificant. The lowest antioxidant activity measured as FRAP was found for WEM extract (6.94±2.59) and GA (1,83±0,69).

3.3.2. Antiradical Activity (A_{AR})

3.3.2.1. DPPH Test

The percent decrease in A_{515} of a methanolic DPPH• solution has been found to be linear in response to increasing amounts of Trolox®, the water-soluble analogue of α-tocopherol. On this basis, A_{AR} of the grape seed extracts could be expressed in terms of Trolox® equivalents (TRE), which appears to be a more descriptive expression, compared with methodologies that calculate only the per cent decrease in A_{515}. Moreover, this approach might be useful in comparing A_{AR} values deriving from other methods using known antioxidants as reference standards. A similar procedure has been used in the estimation of free radical quenching of leaf pigment extracts (Arnao et al. 2001) and wines (Arnous et al. 2002).

A_{AR} of the extracts tested was shown to vary from 0.13 to 1.10 mM TRE/mg GAE, the average being 0.65 mM TRE/mg GAE (Table 3). The order of antiradical efficiency (anti DPPH• radical) of the GSE extracts and GA in decreasing order is:

GA >WEW > WEM>REM>RW>WW (see Table 3).

Table 3. Polyphenolic content and antioxidant properties of the grape seed extracts tested

Extr.	TP(i) (mg GAE/ g seeds)	FRAP (ii) (for 0.3 mg GAE)	A_{AR}		Rancimat (v) I/I_B (hrs)	
			DPPH(iii), EC_{50} (mM Trolox Equivalents /mg GAE)	Chemiluminescence (iv) IC_{50} (mg GAE)	Unpurified oil	Purified oil
RW	40.11 ± 0.90	21.36±3.32[a]	0.99±0.01	67.09±2.18	0.97±0.05[e]	1.75±0.12
WW	38.05 ±0.86	18.28±0.66[a]	1.10±0.03	81.78±0.36	1.02±0.02[e,f]	1.16±0.03
REM	4.58± 0.8	26.90±2.05	0.76±0.04	38.94±0.56	1.09±0.1[f]	1.32±0.06[g]
WEM	0.3±0.01	6.94±2.59[b]	0.25	7.97[d]	1.09±0.10[f]	1.36±0.02[g]
WEW	14.12 ±0.25	151.58± 5.51	0.13[c]	7.12±0.11[d]	1.13±0.04[f]	1.39±0.10[g]
GA	-	1,83±0,69[b]	0.15[c]	4.02 ±0.10	1.1±0.03[f]	2.99±0.13

[a, b, c, d, e, f, g] statistically nonsignificant for $p < 0.05$

IC_{50} Trolox =0.0042mM

Note: Values represent means of at least triplicate determination (n = 3)± S.D. (i) Total phenols expressed as gallic acid equivalents; (ii) reducing power expressed as ascorbic acid equivalents; (iii) anti DPPH activity expressed as Trolox equivalents; (iv) anti hydrogen peroxide action measured by luminol Co(II)-EDTA chemiluminescence; (v) the oxidative stability of sunflower oil (unpurified and purified) given by the induction time.

The highest A_{AR} was found for extract WEW (0.13 mM TRE/mg GAE), having almost the same value like GA (0.15 mM TRE/mg GAE). By contrast, the lowest A_{AR} was observed for extract WW (1.10 mM TRE/mg GAE). Despite its low FRAP value, WEM registers a very good A_{AR} (0.25 mM TRE/mg GAE), followed by REM (0.76±0.04 mM TRE/mg GAE), RW (0.99±0.01 mM TRE/mg GAE) and WW (1.10±0.03 mM TRE/mg GAE).

3.3.2.2. Chemiluminescence

The measurement is based on the ability of each extract to reduce the concentration of hydrogen peroxide, and hence reduce the I_o of the luminol-hydrogen peroxide chemiluminescence reaction. The order of antiradical efficiency (measured as reduction of hydrogen peroxide) of the GSE extracts and GA in decreasing order is:

GA > WEW > WEM>REM>RW>WW (see Table 3).

WW extract has shown the highest IC_{50} value (81.78±0.36) and so being the less antioxidant in terms of hydrogen peroxide scavenging activity. RW, also an aqueous extract but from red grape seeds, registers a higher antioxidant activity, the IC_{50} dropping at 67.09±2.18. The activity of the ethyl acetate extract REM is almost two times higher than that of RW (38.94±0.56 versus 67.09±2.18). An important increase in antioxidant activity towards hydrogen peroxide is measured for WEM and WEW for which the difference is not statistically significant (IC_{50}WEM=7.97 and IC_{50}WEW=7.12±0.11). The highest antioxidant action was registered for the pure GA (IC_{50}=4.02±0.10).

3.3.3. Rancimat

Oxidative stability is an important criterion for evaluating the quality of oils, fats and fatty acid methyl esters (biodiesel). This means that one can now make statements about products stabilised with antioxidants. For example prediction of induction times and the sensible choice of both the type and amount of stabilisers used can be part of the technical specification for suppliers.

The operating principle of the Rancimat method sees a stream of air is blown through the sample at a temperature between 50° to 220° C. This oxidises the fatty acids in several stages. In principle oxidation takes place according to a radical chain mechanism, in which easily volatile oxidation products (chiefly formic acid) are finally formed. These are transferred by the stream of air into a measuring vessel containing deionised water, whose conductivity is continually measured. Plotting conductivity against time produces oxidation curves, whose point of inflection is known as the induction time. The ratio between the induction time of the oil with antioxidant (I) and blank (the oil without antioxidant) (I_B) gives the measure of the activity of the antioxidant. When I/I_B is lower than 1 there is indication of no antioxidant action but of a prooxidant one, and when is higher than 1 the antioxidant action of the compound is shown. The conductivity was measured for the unpurified sunflower oil and for the purified one and the ratios I/I_B are presented in Table 3. The oxidative stability of the unpurified oil in the presence of antioxidants decreased as follows:

WEW>GA> WEM=REM>RW>WW (see Table 3) and of the purified oil as follows: *GA>RW> WEW>WEM>REM>WW* (see Table 3).

For the unpurified oil the highest antioxidant activity was determined for WEW extract (1.13±0.04). The same level of action was found for WW (1.02±0.02), REM (1.09±0.1) and WEM (1.09±0.04), no statistical significant differences being registered. A slight prooxidant activity was measured for RW, I/I_B= 0.97±0.05. GA (1.1±0.03) has shown an antioxidant activity similar to GSE extracts, WW, REM, WEM and WEW. For the purified oil GA had the highest antioxidant action (I/I_B=2.99±0.13). Also all the extracts tested determined a good oxidative stability, acting as antioxidants with the highest value of extracts' I/I_B (1.75±0.12) registered for RW. The antioxidant activity decreased as follows: WEW (I/I_B=1.39±0.1), WEM (I/I_B=1.36±0.02), REM (I/I_B=1.32±0.06) and WW (I/I_B=1.16±0.03). The unpurified oil contains tocopherols, the difference in conductivity between the unpurified and purified oil being given by the action of these components. It is observed so, an antagonistic action between the oil tocopherols and the extracted polyphenols.

This finding is in concordance with those of Hras et al. (2000), Banias et al. (1992), Peyrat-Maillard et al. (2003) and Samotyja and Malecka (2007). These authors observed antagonism between α-tocopherol and rosmarinic acid or caffeic acid (Peyrat-Maillard et al. 2003; Samotyja and Malecka 2007), between plant extracts rich in polyphenols and α-tocopherol in lard (Banias et al.1992) or sunflower oil (Hras et al. 2000).

3.3.4. Correlation between TP Content of GSEs and Their Antioxidant Activity

The purpose of our extraction system was to use water as the cheapest and most ecological solvent possible. Because than we wanted to see if the sedimentation residue still contains high amounts of polyphenols we performed also an EtOAc extraction. Total polyphenol content was determined for RW, WW, REM, WEM and WEW extracts and decreases in the following order:

RW> WW> WEW> REM> WEM (see Table 3).

The values of the different antioxidant parameters measured, FRAP, A_{AR} and oxidative stability of sunflower oil, were plotted against the TP values and so the degree of correlation between the total polyphenol content measured by Folin-Ciocalteu method and the antioxidant properties of the studied extracts was determined (Table 4). The degrees of correlation between the different antioxidant parameters were calculated and they are presented also in Table 4.

From Table 4 it can be seen that no correlation was found between the TP content of the extracts and FRAP (r^2=0.02) and the oxidative stability of the purified oil containing the antioxidants determined by Rancimat (r^2=0.09). A better correlation has the TP content with A_{AR} activity in case of DPPH• radical (r^2=0.56), even better with the oxidative stability of the unpurified sunflower oil containing the antioxidants (r^2=0.72) and the best was registered for A_{AR} activity determined by chemiluminescence measurements. No correlation is observed between FRAP and the other antioxidant measurement methods used in this study.

Table 4. Square correlation coefficient (r^2) calculated for the TP content of the tested extracts (RW, WW, REM, WEM, WEW) and for their different antioxidant parameters

		TP	FRAP	A_{AR}		Rancimat	
				DPPH	Chemil	Unpurified oil	Purified oil
TP							
FRAP		0.02	-				
A_{AR}	DPPH	0.56	0.13	-			
	Chemil	0.72	0.08	0.97	-		
Rancimat	Unpurified oil	0.70	0.20	0.86	0.74	-	
	Purified oil	0.09	0.09	0.18	0.17	0.02	-

Also for the oxidative stability of the sunflower oil containing antioxidants no correlation was found with the other methods, not even with the stability of the unpurified oil. Surprisingly good correlation was determined between the oxidative stability of the unpurified oil and the A_{AR} activity, with r^2=0.74 in case of chemiluminescence and r^2=0.86 in case of DPPH• radical. As expected the two antiradical methods the DPPH• test and chemiluminescence are very highly correlated (r^2=0.97). When A_{AR} values were plotted against TP content, a r^2 = 0.7082 was calculated, indicating that the antiradical efficiency of the wines may be linked to their TP concentration (Arnous et al. 2002). This finding provided evidence that A_{AR} might be significantly associated with flavanols (catechin, epicatechin, proanthocyanidins) (Arnous et al. 2002).

CONCLUSIONS

In grape seeds, procyanidins represent in general the major part of the total polyphenol extract, and their extreme complexity is the result of the large number of different compounds with very similar structures. By LC-DAD-MS in ESI+ mode we have identified in the aqueous extracts RW and WW, gallic acid, catechin, epicatechin, procyanidine dimers, trimers, tetramers, pentamers and hexamers. Procyanidin oligomers with a higher degree of galloylation were identified when the sedimentation residues from the first extraction were reextracted in ethyl acetate. For REM and WEM extracts the major peaks eluting at 38 min were tentatively attributed to isomeric structures of a procyanidine nanomer trigalloylated. For WEW extract the major peak would show the elution of a procyanidin octamer trigallate.

Different extraction solvents determined the extraction of different procyanidine molecules besides the monomers catechin and epicatechin. Using hot water, procyanidin oligomers with not a high degree of galloylation were extracted. Extracting the second time with EtOAc, we get procyanidins with a higher galloylation degree which also influence their antioxidant properties.

The highest content in total polyphenols was registered for the aqueous extracts but this was not necessarily linked to a higher antioxidant activity. A very high ferric reducing power was measured for the aqueous phase of the EtOAc extract (WEW). This extract showed also the highest activity in terms of antiradical power and protection of the unpurified oil oxidation. The EtOAc extracts proved to have the highest action compared to the aqueous ones, as antiradical agents, measured against the DPPH and hydrogen peroxide. It seems that the galloylation difference between RW and WW, increased in the case of RW, doesn't make any difference for FRAP. WEW had the highest FRAP followed by REM. In case of REM extract, even though the TP is lower than RW and WW, its FRAP shows a higher value. In this case the galloylation seems to have a great influence in determining FRAP.

The oxidation stability of the unpurified and purified sunflower oil in the presence of the antioxidant extracts measured through the Rancimat method have shown a strong interaction between the polyphenols and tocopherols. According to the results of this study the tocopherols from the unpurified oil have an antagonistic action against the polyphenolic extracts, reducing the antioxidant activity of those. This fact can be seen very well following the behaviour of the aqueous extract form the red grape seeds (RW) added to the unpurified and then to the purified oil. For the unpurified oil all the extracts had almost the same antioxidant activity statistically no different from the one of the pure GA, but for RW extract a slight prooxidant action was registered. The situation changes completely when the tocopherols were removed from the oil, from all the extracts RW induces the highest oxidation stability for the purified oil.

We can conclude that besides the fact that due to the environmental and each grape cultivar's genetic inheritance influence, a specific antioxidant polyphenolic "print" can be analyzed, the mixture of different cultivars can also provide unique "recipes" for polyphenolic GSE with desired antioxidant action through the valorization of waste products of the winery and grape juice industry.

ACKNOWLEDGEMENTS

V. S. Chedea was a fellow of the Romanian Ministry of Research, Education and Innovation. The postdoc fellowship was conferred accordingly to H.G. nr. 697/1996, modified by H.G. nr. 533/1998.

REFERENCES

Amico, V., Napoli, E.M., Renda, A., Ruberto, G., Spatafora, C., Tringali C. (2004). Constituents of grape pomace from the Sicilian cultivar "Nerello Mascalese". *Food Chemistry 88*, 599-607.

Arnao, M. B., Cano, A., Alcolea, J. F., Acosta, M. (2001). Estimation of free radical-quenching activity of leaf pigment extracts. *Phytochemical Analysis, 12*, 138-143.

Arnous, A., Makris, D. P., Kefalas, P. (2001). Effect of principal polyphenolic components in relation to antioxidant characteristics of aged red wines. *Journal of Agricultural and Food Chemistry, 49,* 5736–5742.

Arnous, A., Makris, D. P., Kefalas, P. (2002). Correlation of pigment and flavanol content with antioxidant properties in selected aged regional wines from Greece. *Journal of Food Composition and Analysis, 15*, 655–665.

Balu M., Sangeetha P., Haripriya D., Panneerselvam C. (2005), Rejuvenation of antioxidant system in central nervous system of aged rats by grape seed extract, *Neuroscience Letters, 383*, 295-300.

Banias, C., Oreopoulou, V., Thomopoulos, C.D. (1992). The effects of primary antioxidants and synergists on the activity of plant-extracts in lard. *Journal of American Oil Chemistry Society, 69*, 520–524.

Benzie, I.F.F., Strain, J.J. (1996) The ferric reducing ability of plasma (FRAP) as a measure of "antioxidant power": The FRAP assay. *Analytical Biochemistry 239*, 70–76.

Chedea, V.S., Echim, C., Braicu, C. Andjelkovic, M., Verhe, R., Socaciu, C. (2011). Composition in polyphenols and stability of the aqueous grape seed extract from the Romanian variety "Merlot Recas". *Journal of Food Biochemistry 35*, 92-108.

De Freitas, V. A. P., Glories, Y., Bourgeois, G., Vitry, C. (1998). Characterisation of oligomeric and polymeric procyanidins from grape seeds by Liquid Secondary Ion Mass Spectrometry. *Phytochemistry, 38*, 1435-1441.

El-Ashmawy, I. M., Gad, S. B., Salama, O. M. (2010). Grape seed extract prevents azathioprine toxicity in rats. *Phytotherapy Research, DOI: 10.1002/ptr.3200.*

Fuleki, T., Ricardo da Silva, J.M., (1997). Catechin and procyanidin composition of seeds from grape cultivars grown in Ontario. *Journal of Agriculture and Food Chemistry, 45*, 1156–1160.

Gomez-Alonso, S., Garcia-Romero, E., Hermosin-Guitierrez, I. (2007). HPLC analysis of diverse grape and wine phenolics using direct injection and multidetection by DAD and fluorescence. *Journal of Food Composition and Analysis, 20*, 618-626.

Gonzales-Manzano, S., Rivas-Gonzalo, J.C., Santos-Buelga, C. (2004). Extraction of flavan-3-ols from grape seed and skin into wine using simulated maceration. *Analytica Chimica Acta, 513*, 283-289.

Guendez, R., Kallithraka, S., Makris, D. P., Kefalas, P. (2005). An analytical survey of the polyphenols of seeds of varieties of grape (*Vitis vinifera*) cultivated in Greece: Implications for exploitation as a source of value-added phytochemicals, *Phytochemical Analysis, 16*, 17–23.

Hayasaka, Y., Waters, E. J., Cheynier, V., Herderich, M. J., Vidal, S. (2003). Characterization of proanthocyanidins in grape seeds using electrospray mass spectrometry. *Rapid Communications in Mass Spectrometry, 17*, 9-16.

Hras, A.R., Hadolin, M., Knez, Z., Bauman, D. (2000). Comparison of antioxidative and synergistic effects of rosemary extract with alpha-tocopherol, ascorbyl palmitate and citric acid in sunflower oil. *Food Chemistry, 71*, 229–33.

Jordão, A.M., Ricardo-da-Sillva, J.M., Laureano, O. (2001). Evolution of catechins and oligomeric procyanidins during grape maturation of Castelão Francês and Touriga Francesa. *American Journal of Enology and Viticulture, 52*, 230-237.

Mateus, N., Marques, S., Gonçalves, A.C., Machado, J.M., de Freitas, V.A.P. (2001). Proanthocyanidin composition of red Vitis vinifera varieties from the Douro valley during ripening: influence of cultivation altitude. *American Journal of Enology and Viticulture, 52*, 115-124.

Morthup, R. R., Dahlgren, R. A., Mccoll, J. G. (1998). Polyphenols as regulators of plant-litter-soil interactions in northern California's pygmy forest: A positive feedback? *Biogeochemistry, 42*, 189–220.

Núñez, V., Gomez-Cordoves, C., Bartolome, B., Hong, Y.-J., Mitchell, A. E. (2006). Non-galloylated and galloylated proanthocyanidin oligomers in grape seeds from *Vitus vinifera* L. cv. Graciano, Tempranillo and Cabernet Sauvignon. *Journal of the Science of Food and Agriculture 86*, 915–921.

Oszmianski, J., Sapis, J.C. (1989) Fractionation and identification of some low molecular weight grape seed phenolics. *Journal of Agriculture and Food Chemistry, 37*, 1293–1297.

Parejo, I., Codina, C., Petrakis, C., Kefalas, P. (2000). Evaluation of scavenging activity assessed by Co(II)/EDTA-induced luminol chemiluminescence and DPPH (2,2 diphenyl-1-picrylhydrazyl) free radical assay. *Journal of Pharmacological and Toxicological Methods 44*, 512–597.

Pati, S., Losito, I., Gambacorta, G., La Notte, E., Palmisano, F., Zambonin, P. G. (2006). Simultaneous separation and identification of oligomeric procyanidins and anthocyanin-derived pigments in raw red wine by HPLC-UV-ESI-MS^n. *Journal of Mass Spectrometry, 41*, 861–871.

Peyrat-Maillard, M.N., Cuvelier, M.E., Berset, C. (2003). Antioxidant activity of phenolic compounds in 2,2_-azobis (2-amidinopropane) dihydrochloride (AAPH)-induced oxidation: synergistic and antagonistic effect. *Journal of American Oil Chemical Society, 80*, 1007–1012.

Prasain, J. K., Peng, N., Dai, Y., Moore, R., Arabshahi, A., Wilson, L., Barnes, S., Wyss, J. M., Kim, H., Watts, R. L. (2009), Liquid chromatography tandem mass spectrometry identification of proanthocyanidins in rat plasma after oral administration of grape seed extract. *Phytomedicine, 16,* 233-243.

Rubilar, M., Pinelo, M., Shene, C., Sineiro, J., Nuñez, M. J. (2007). Separation and HPLC-MS Identification of Phenolic Antioxidants from Agricultural Residues: Almond Hulls and Grape Pomace. *Journal of Agricultural and Food Chemistry, 55,* 10101–10109.

Santos-Buelga, C., Francia-Aricha, M.E., Escribano-Bailón, M.T. (1995). Comparative flavan-3-ol composition of seeds from different grape varieties. *Food Chemistry, 53*, 197–201.

Samotyja, U., Malecka, M. (2007). Effects of blackcurrant seeds and rosemary extracts on oxidative stability of bulk and emulsified lipid substrates. *Food Chemistry, 104*, 317–323.

Schewe, T., Kühn, H., Sies, H. (2002). Flavonoids of cocoa inhibit recombinant human 5-lipoxygenase. *Journal of Nutrition, 132*, 1825-1829.

Schewe, T., Sadik, C., Klotz, L.-O., Yoshimoto, T., Kühn, H., Sies, H. (2001). Polyphenols of cocoa: inhibition of mammalian 15-Lipoxygenase. *Biological Chemistry, 382*, 1687 – 1696.

Shi, J., Yu, J., Pohorly, J.E., Kakuda, Y. (2003). Polyphenolics in grape seeds-Biochemistry and functionality. *Journal of Medicinal Food, 6*, 291–299.

Sun, B., Belchior, G.P., Ricardo da Silva, J.M., Spranger, I. (1999). Isolation and purification of dimeric and trimeric procyanidins form grape seeds. *Journal of Chromatography A, 84*, 1115–121.

Surh,Y.J. (2003). Cancer chemoprevention with dietary phytochemicals. *Nature Reviews Cancer, 3*, 768-780.

Tong, H., Bell, D., Tabei, K., Siegel, M.M. (1999). Automated data messaging, interpretation, and e-mailing modules for high throughput open access mass spectrometry. *Journal of American Society of Mass Spectrometry, 10*, 1174–1187.

Torres, J. L., Varela, B., García, M. T., Carilla, J., Matito, C., Centelles, J. J., Cascante, M., Sort, X., Bobet, R. (2002). Valorization of grape (*Vitis vinifera*) byproducts. Antioxidant and biological properties of polyphenolic fractions differing in procyanidin composition and flavonol content. *Journal of Agriculture and Food Chemistry, 50*, 7548 -7555.

Wu, T.-H., Liao, J.-H.,. Hsu, F.-L, Wu, H.-R., Shen, C.-K., Yuann, J.-M. P., Chen, S.-T. (2010). Grape seed proanthocyanidin extract chelates iron and attenuates the toxic effects of 6-hydroxydopamine: implications for Parkinson's disease. *Journal of Food Biochemistry, 34*, 244-262.

Xie, Q., Bedran-Russo, A. K., Wu, C. D. (2008). *In vitro* remineralization effects of grape seed extract on artificial root caries, *Journal of Dentistry, 36*, 900-906.

Yang, Y., Chien, M. (2000). Characterization of grape procyanidins using High-Performance Liquid Chromatography/Mass Spectrometry and Matrix-Assisted Laser Desorption/Ionization Time-of-Flight Mass Spectrometry, *Journal of Agriculture and Food Chemistry 48,* 3990-3996.

In: Grapes ISBN 978-1-61470-950-3
Editors: R. P. Murphy et al., pp. 107-132

Chapter 4

PHYSIOLOGICAL AND MOLECULAR CHARACTERISTICS OF JAPANESE INDIGENOUS *VITIS VINIFERA* CV. KOSHU GRAPE

***Shunji Suzuki*[a] *and Hironori Kobayashi*[b]**
[a]Laboratory of Fruit Genetic Engineering,
The Institute of Enology and Viticulture, University of Yamanashi,
Kofu 400-0005, Japan
[b]Château Mercian, Mercian Corporation, Katsunuma,
Koshu 409-1313, Japan

ABSTRACT

Koshu (*Vitis vinifera* cv.) is an indigenous grape cultivar that has been grown for more than a thousand years in Japan. Koshu has been categorized in the cluster of oriental cultivars of *V. vinifera*. Yamanashi Prefecture accounts for more than 90% of Koshu grape production in Japan. To improve varietal aroma and taste of wine made from Koshu grapes, various enological techniques, such as sur lies, cryo-extraction, skin contact, and barrel fermentation, have been used by several wineries. Koshu grapes have certain characteristics that are exemplified by large berry size, pink skin color, and high total phenolic content, which differ from those of European grapes generally used for winemaking. However, physiological and molecular studies of Koshu grape are lacking. In this review, we present the physiological and molecular characteristics of Koshu grape, which have been clarified from our studies aimed at improving Koshu grape and wine quality. This chapter is expected to contribute to a better understanding of this indigenous grape cultivar and its products worldwide.

INTRODUCTION

Vitis vinifera cv. Koshu grapes (Figure 1a) are widely cultivated in Central Japan, particularly Yamanashi Prefecture. Koshu grape is an indigenous grape cultivar that has been grown for more than a thousand years in Japan. Today, it is one of the most abundantly produced grape varieties in Japan. Koshu grape has been categorized in the cluster of oriental cultivars of *V. vinifera* (Figure 1b) (Goto-Yamamoto et al., 1998, 2006). The grapes are typically used in making Japanese high-quality wines. However, Koshu grapes have certain characteristics that differ from those of European grapes (*V. vinifera* L.) generally used for winemaking. For example, fresh berry weight of Koshu grapes at the time of harvest is more than twice those of Chardonnay, Sauvignon Blanc, Merlot, and Cabernet Sauvignon grapes, and Koshu grape skin becomes light purple during growth (Figure 1a). Moreover, Koshu grape juice has higher total phenolic (TP) content than juice from other white grape cultivars, and wine made from Koshu grapes also has higher TP content than white wine made from other European grape cultivars. Okamura and Watanabe (1981) have reported that the contents of caftaric and coutaric acids, which are the tartrate derivatives of caffeic and coumaric acids, respectively, are much higher in Koshu wines than in white wines made from Semillon, Chardonnay, and Riesling grapes. The high contents of these compounds are considered to affect the taste of Koshu wine (Yokotsuka, 1995). In addition to grape characteristics, the cultivation conditions of Koshu grapes differ from those of European grapes. In general, numerous Koshu grape vineyards demonstrate not guyot-style but shelf-style cultivation (overhead trellis) (Figure 1c).

To produce high-quality Koshu wine, various enological techniques, such as the improvement of amino acid content using *sur lies* (Ari'Izumi et al., 1994), the amelioration of non-flavonoid compounds by barrel fermentation (Yokotsuka et al., 1994), the removal of bitterness by hyperoxidation (Yokotsuka et al., 2005), and the extraction of aromatic precursors by skin contact (Kobayashi et al., 2007), have been demonstrated.

(a) (c)

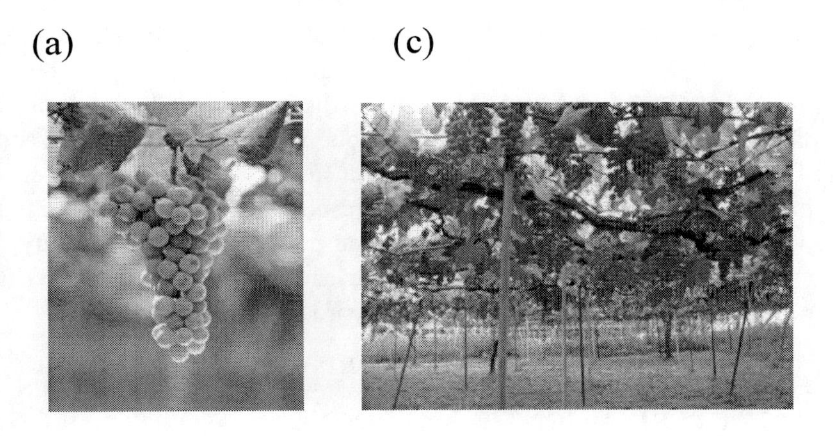

(b)

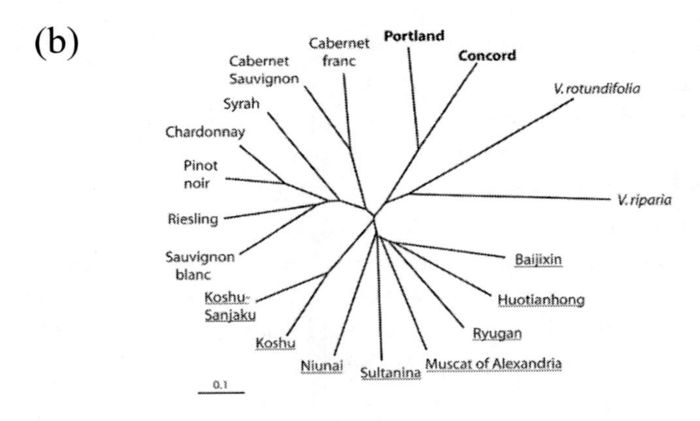

Figure adapted from Goto-Yamamoto et al. (2006).

Figure 1. Japanese indigenous grape cultivar ‘Koshu’. (a) Outer appearance of ‘Koshu’. (b) UPGMA dendrogram based on the phenetic distances of the tested grape samples. Occidental and Oriental cultivars of *V. vinifera* are shown by Roman font and underlined Roman font, respectively. Bold font represents the cultivar of *V. labrusca*. Figure adapted from Goto-Yamamoto et al. (2006). (c) Shelf-style cultivation (overhead trellis).

Those enological techniques have contributed to the establishment of the original style and viticultural practices of Koshu wines.

In this review, to present the physiological and molecular characteristics of Koshu grape, we summarized 1) anthocyanin composition and accumulation in Koshu grape skin, 2) the distribution of hydroxycinnamic acids, monomeric flavan-3-ols, and flavonols, in Koshu grape tissue, and 3) the evolution of flavor precursors in Koshu grape berries grown in several regions in Japan, during development. The results are expected to contribute to the improvement of Koshu grape viticultural practices and wine quality, and may be useful for the modification of the chemical composition of Koshu grape and wine using genetic engineering approaches,

CHARACTERIZATION OF ANTHOCYANIN COMPOSITION IN KOSHU GRAPE SKIN

Skin anthocyanin content is responsible for the black/red color of grape berries (Glories, 1978). Boss et al. (1996) have reported that anthocyanins are synthesized from phenylalanine and that anthocyanin pigments form glucoside conjugates to improve their stability (Figure 2).

Anthocyanin composition is also a key contributor to wine color (Castellarin et al., 2006; Castellarin and Gaspero, 2007). In addition to their role in coloration, the anthocyanin synthetic pathway relates a wide range of compounds, such as phenylpropanoids, stilbenes, flavan-3-ols, and flavonols, in many plants and fruit, including grapes (Bogs et al., 2006; Jeong et al., 2006). Therefore, the composition and content of anthocyanins in grape berry affect wine taste (Glories, 1988) as well as wine color. Kobayashi et al. (2001) have reported that the coloration in grape berries is determined by the activity of UDP-glucose:flavonoid 3-O-glucosyltransferase (UFGT). *UFGT* gene expression is regulated by VvmybA, which acts as a transcription factor (Kobayashi et al., 2002). The insertion of *Gret1* retrotransposon in *VvmybA* gene is the molecular basis of white-colored grape berries (Kobayashi et al., 2004b; 2005). This phenomenon reduces UFGT expression in green-yellow-colored grape berries and as a result, anthocyanins are hardly accumulated in green-yellow-colored grape skin (Furiya et al., 2009; Kobayashi et al., 2005; Walker et al., 2007).

Anthocyanins are classified into five groups: cyanidin, peonidin, delphinidin, petunidin, and malvidin, based on the number and position of hydroxyl groups on the B-ring. Cyanidin and peonidin are called cyanidin-based anthocyanins because they are synthesized from cyanidin.

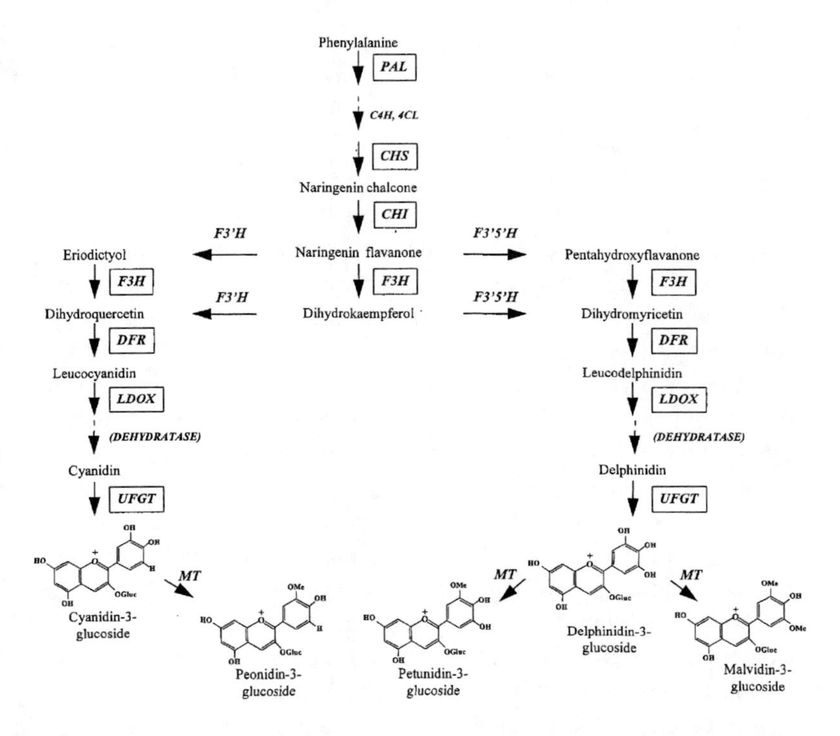

Figure 2. Simplified schematic of the anthocyanin biosynthetic pathway, modified to account for the major products found in grapes. PAL, phenylalanine ammonia lyase; C4H, cinnamate 4-hydroxylase; 4CL, 4-coumarate ligase; CHS, chalcone synthase; CHI, chalcone isomerase; F3H, flavanone-3-hydroxylase; F3'H, flavonoid 3'-hydroxylase; F3',5'H, flavonoid 3',5'-hydroxylase; DFR, dihydroflavonol 4-reductase; LDOX, leucoanthocyanidin dioxygenase; UFGT: UDP glucose-flavonoid 3-o-glucosyl transferase; MT, methyltransferase. Figure adapted from Boss et al. (1996).

Delphinidin, petunidin, and malvidin are delphinidin-based anthocyanins because they are synthesized from delphinidin (Nyman and Kumpulainen 2001). Anthocyanins are sensitive to pH, changing from red to blue as the pH rises. In grape skin that generally has a pH of 4 or lower, cyanidin-based anthocyanins produce a red color, whereas delphinidin-based anthocyanins produce a blue color. The composition of anthocyanins (ratio of cyanidin- to delphinidin-based anthocyanins) in berry skin has a marked effect on the expression of flavonoid 3'-hydroxylase (*F3'H*) and flavonoid 3',5'-hydroxylase (*F3',5'H*) (Figure 2, Castellarin et al., 2006). Therefore, the transcriptional induction of *F3'H* and *F3',5'H* coincides with the beginning of anthocyanin biosynthesis in black-colored berry and green-yellow-colored berry (Castellarin et al., 2006). In addition, *F3'H* and *F3',5'H* expression is consistent with the chromatic evolution of ripening berries (Castellarin et al., 2006), and anthocyanin composition in berry skin differs among grape cultivars. Thus, the accumulation ratio of cyanidin-based anthocyanins to delphinidin-based ones in grape berry may be determined by the transcription patterns of *F3',5'H* and *F3'H* genes during development. Moreover, certain environmental conditions affect anthocyanin content and composition via the expression of *F3'H* and/or *F3',5'H*, such as solar radiation (Cortell and Kenndy, 2006), air temperature (Mori et al., 2007; Cohen et al., 2008), and water deficit (Castellarin et al., 2007).

Koshu grape berry has light purple skin and anthocyanin content increases during development (Figure 3a). During berry development, cyanidin-based anthocyanins accumulate in berry skin, accounting for approximately 30-60% of the total anthocyanins accumulated in skin at the end of the ripening process. Black-colored Merlot grape skin accumulates more delphinidin-based anthocyanins than cyanidin-based anthocyanins, whereas the reverse is true for green-yellow-colored Chardonnay grape skin. The ratio of cyanidin-based anthocyanins to delphinidin-based ones in Koshu grape skin is intermediate between those in Merlot and Chardonnay grape skins (Figure 3b). Delphinidin-based and cyanidin-based anthocyanin composition ratio in each grape cultivar can be explained by *F3'H* and *F3',5'H* gene transcription analysis (Figure 3c). In addition, anthocyanin content in each grape skin can be explained by *UFGT* gene transcription analysis as well as anthocyanin composition ratio (Figure 3d). From the results, the factors responsible for the formation of light purple Koshu grape skin have been established (Kobayashi et al., 2009).

Characterization of Phenolic Compounds Biosynthesized in Koshu Grape

Grape berry contains various phenolic compounds, such as anthocyanins, stilbenes, hydroxycinnamic acids, phenylpropanoids, flavan-3-ols, proanthocyanidins, and flavonols. The distribution of phenolic compounds differs among the grape cultivars and influences the individuality of each grape. In this chapter, we explain the distribution of various phenolic compounds in Koshu grape tissue compared with other grape cultivars.

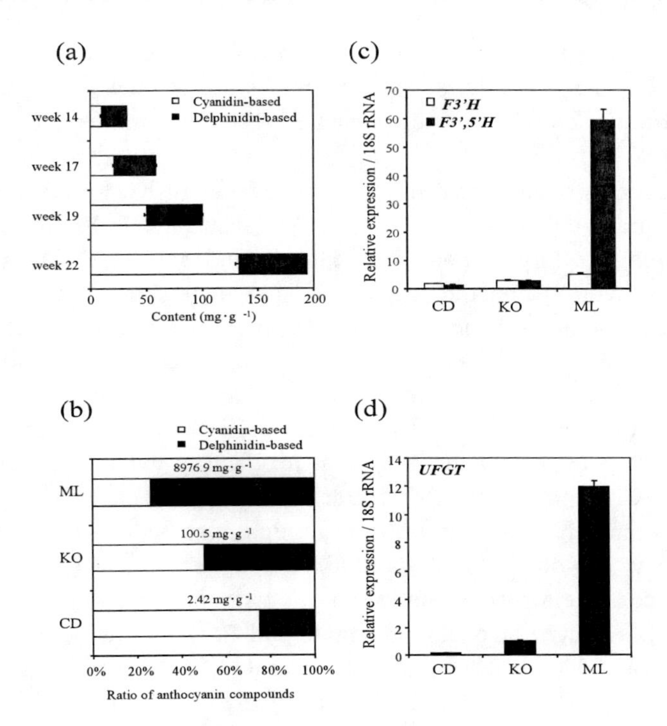

Figure 3. Anthocyanin in 'Koshu' berry. (a) Anthocyanin composition in berry skin of light purple Koshu grapes during development. Week 14: 14 weeks post flowering at the end of véraison, week 17: 17 weeks post flowering at early harvest, week 19: 19 weeks post flowering at harvest, week 22: 22 weeks post flowering at the end of the ripening process. White column indicates cyanidin-based anthocyanins (cyanidin and peonidin). Black column indicates delphinidin-based anthocyanins (delphinidin, petunidin, and malvidin). Data represent the results of triplicate experiments. (b) Anthocyanin composition in berry skin of different cultivars. Merlot (ML), Koshu (KO), and Chardonnay (CD) at 19 weeks post flowering at harvest. White column indicates cyanidin-based anthocyanins (cyanidin and peonidin). Black column indicates delphinidin-based anthocyanins (delphinidin, petunidin, and malvidin). Values in figure indicate total amount of anthocyanins in one gram of fresh weight of each tissue. (c) *F3'H* and *F3',5'H* expression in berry skin of different cultivars. White column indicates *F3'H*. Black column indicates *F3',5'H*. *F3'H* and *F3',5'H* expression was determined by real-time RT-PCR. 18S rRNA was used as internal control. Data were calculated as gene expression relative to 18S rRNA gene expression. Bars indicate means ± standard deviation of triplicate experiments. (d) *UFGT* expression in berry skin of different cultivars. *UFGT* expression was determined by real-time RT-PCR. 18S rRNA was used as internal control. Data were calculated as gene expression relative to 18S rRNA gene expression. Bars indicate means ± standard deviation of triplicate experiments. Data adapted from Kobayashi et al. (2009).

Koshu grape juice has higher total phenolic content than the juices of other white grape cultivars, and Koshu wines are frequently singled out for their bitterness and/or astringency compared with other white wines.

In addition, the accumulation of hydroxycinnamic acids in Koshu grape skin at harvest is one of the distinct characteristics of Koshu grape compared with other grape cultivars (Okamura and Watanabe, 1981; Yokotsuka, 1995). For example, the large amounts of caftaric and coutaric acids, which are examples of hydroxycinnamic acids, in Koshu grape skin and pulp contribute to the large amounts of phenols in Koshu juice that was extracted by high-pressure pressing (Yokotsuka, 1990). Coutaric acid is synthesized by conjugating coumaric acid with tartaric acid. In grape berry, coumaric acid acts as a flavor precursor of volatile phenols, including 4-vinylphenol, a compound responsible for the well-known olfactory default called 'phenolic off-flavor' (Chatonnet et al., 1993). Our previous study has demonstrated that almost all commercial Koshu wines have high contents of 4-vinylphenol and 4-vinylguaiacol (Kobayashi et al., 2006) and suggested that those compounds may be responsible for the phenolic and smoky aroma in Koshu wines.

The gross weight of phenylpropanoid and flavonol compounds in Koshu grape skin is 1.84-fold, 1.91-fold, and 1.43-fold of those in Sauvignon Blanc, Chardonnay, and Merlot grape skins, respectively (Table 1).

Koshu grape skin contains 639.2 μg of *p*-coutaric acid per g of fresh weight, which is the highest among the grape skins tested (4.76-fold of Sauvignon Blanc; 4.19-fold of Chardonnay; 9.52-fold of Merlot) (Figure 4a). Transcription profiles are summarized in Figure 5.

Cinnamic acid 4-hydroxylase (*C4H*), *p*-coumarate 3-hydroxylase (*C3H*), and caffeate methyltransferase (*CMT*), which are related to the biosynthesis of hydroxycinnamic acids, are abundantly expressed in Koshu grape skin compared with the other grape skins (Figures 5a, b, and d). Chemical and transcription analyses suggest that Koshu grape berry skin has high hydroxycinnamic acid content and that the skin contact and/or maceration methods used in Koshu winemaking may extract large amounts of hydroxycinnamic acids from skin into juice, resulting in the strong astringency of Koshu wines.

Flavonols act as a stabilizer of wine color by copigmentation with anthocyanins (Baranac et al., 1997) and most of them form glucoside conjugates (Cheynier and Rigaud, 1986). Similar to hydroxycinnamic acids, flavonol accumulation in Koshu skin is higher than those in the other grape cultivars (Table 1).

Table 1. Total phenolic, anthocyanin, proanthocyanin, and other phenolic contents in grape tissues. Data adapted from Kobayashi et al. (2011a)

Tissue	Cultivar	Total phenolics [a]	Total anthocyanins [a]	Total proanthocyanidins [a]	Phenylpropanoids and flavonols [a]
Skin	SB [b]	7033 (±75.2) [c]	283.5 (±2.9)	1593 (±78.6)	5157 (±33.3)
	CD [b]	6183 (±50.1)	218.8 (±16.8)	989 (±57.1)	4976 (±61.9)
	KO [b]	12300 (±102.1)	344.1 (±14.3)	2444 (±34.9)	9511 (±101.9)
	ML [b]	19648 (±241.4)	10357.1 (±178.9)	2675 (±56.1)	6616 (±82.7)
Pulp	SB	606.7 (±12.6)	76.5 (±0.7)	66.8 (±8.1)	463 (±10.9)
	CD	696.7 (±28.4)	113.4 (±2.9)	114.2 (±6.2)	469 (±22.1)
	KO	823.3 (±27.5)	106.3 (±3.4)	117.9 (±11.3)	599 (±20.6)
	ML	880.1 (±44.4)	185.6 (±6.6)	122.9 (±6.1)	572 (±48.1)
Seed	SB	72532 (±608.9)	352.9 (±27.1)	56791 (±1796.5)	15388 (±435.1)
	CD	58612 (±517.3)	253.8 (±29.5)	41890 (±1590.8)	16468 (±112.5)
	KO	85266 (±1035.6)	335.7 (±25.1)	47965 (±1749.1)	36965 (±645.6)
	ML	110832 (±1818.7)	323.1 (±21.8)	80070 (±4425.1)	30439 (±604.5)

[a] mg/g of fresh weight
[b] SB, Sauvignon Blanc; CD, Chardonnay; KO, Koshu; and ML, Merlot
[c] Data are shown as means ± standard deviations of triplicate experiments from three independent samples..

In the case of flavonols (Figure 4b), quercetin-3-O-galactoside content in Koshu berry skin is 92.7 μg/g of fresh weight, the highest among the grape skins tested (2.34-fold of Sauvignon Blanc; 1.30-fold of Chardonnay; 2.13-fold of Merlot). In addition, Koshu grape skin accumulates more quercetin-3-O-glucoside than the other grape skins, the content being 500.6 μg/g of fresh weight (1.45-fold of Sauvignon Blanc; 1.39-fold of Chardonnay; 1.80-fold of Merlot), and quercetin-3-O-glucuronide content in Koshu grape skin is also higher than those in the other grape skins, being 330.5 μg/g of fresh weight (2.11-fold of Sauvignon Blanc; 5.63-fold of Chardonnay; 1.78-fold of Merlot). Flavonol synthase is the key enzyme for the conversion of dihydroflavonols into flavonols. Grape has five genes that encode flavonol synthases, *VvFLS1* to *VvFLS5* (Fujita, 2008). *VvFLS1*, *VvFLS2*, and *VvFLS3* are mainly expressed in leaf, bud, and inflorescence, but not in grape skin at harvest (Downey et al., 2003, Fujita, 2008).

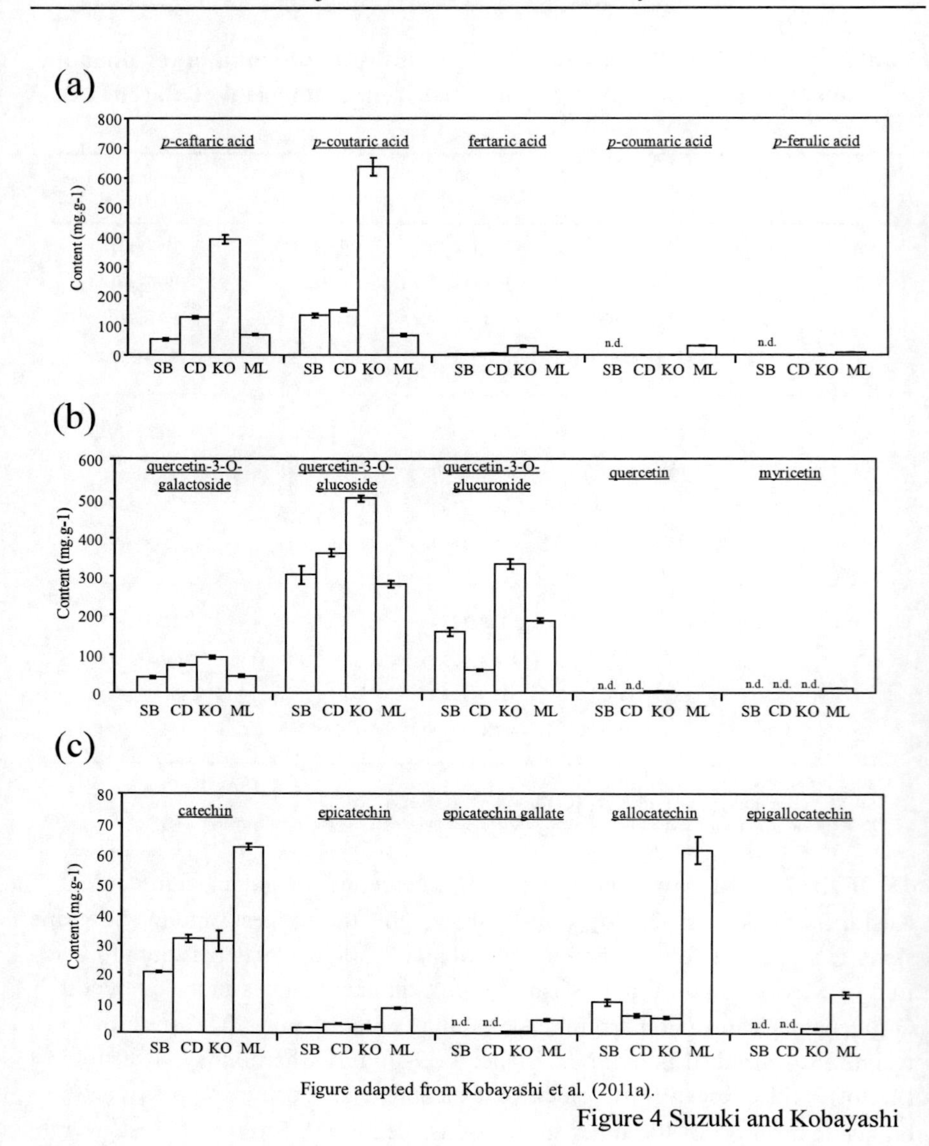

Figure 4. Hydroxycinnamic acid, monomeric and dimeric flavan-3-ol, and monomeric flavonoid contents in grape skin of four cultivars. Grape berries were collected at 19 wpf in 2009 growing season. SB, Sauvignon Blanc; CD, Chardonnay; KO, Koshu; and ML, Merlot. n.d., not detected. Data are shown as means ± standard deviation of triplicate experiments from three independent samples. Data adapted from Kobayashi et al. (2011a).

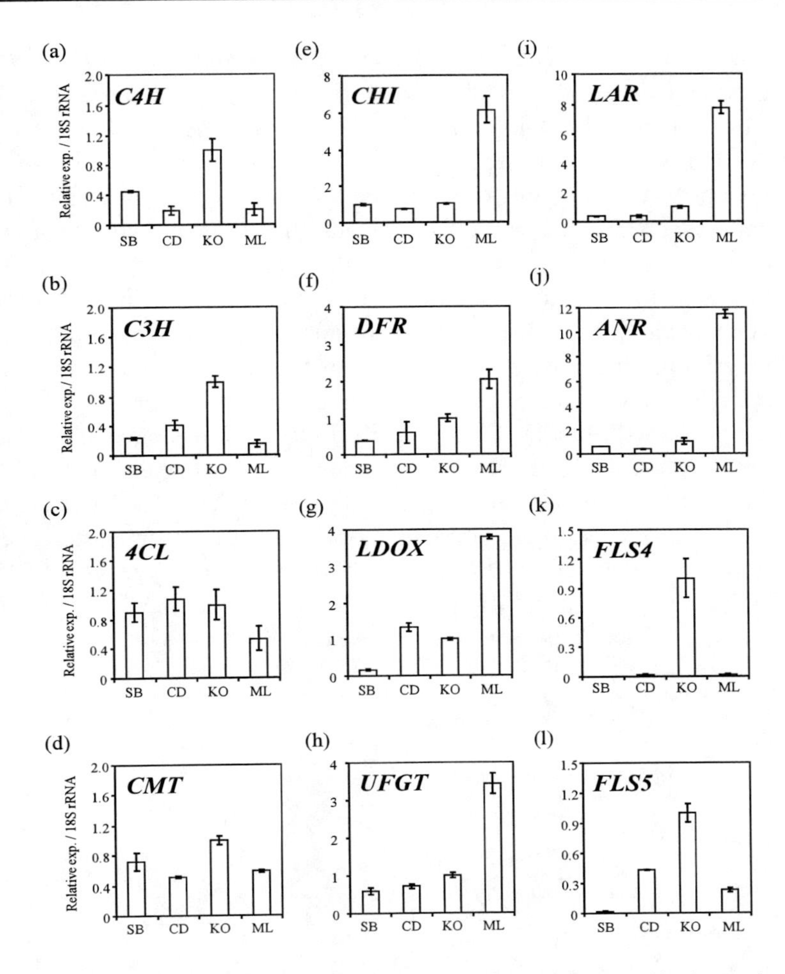

Figure 5. Transcription analysis of genes related to the synthesis of hydroxycinnamic acids, flavan-3-ols, and flavonoids in Koshu grape skin. C4H, Cinnamic acid 4-hydroxylase; C3H, *p*-coumarate 3-hydroxylase; 4CL, 4-coumaroyl CoA ligase; CMT, caffeate methyltransferase; CHI, chalcone isomerase; DFR, dihydroflavonol 4-reductase; LDOX, leucoanthocyanidin dioxygenase; UFGT, UDP glucose-flavonoid 3-O-glucosyl transferase; LAR, leucoanthocyanidin reductase; ANR, anthocyanidin reductase; FLS, flavonol synthase (a-l). Transcription profiles of the genes. 18S rRNA was used as internal control. SB, Sauvignon Blanc; CD, Chardonnay; KO, Koshu; and ML, Merlot. Data were calculated as gene expression relative to 18S rRNA expression, and gene expression was normalized to that of KO. Bars indicate means ± standard deviation of triplicate experiments from three independent samples. Data adapted from Kobayashi et al. (2011a).

Our transcription analysis of *FLS* genes is consistent with those reports, demonstrating that Koshu grape skin expresses *FLS4* and *FLS5* genes at harvest (Figures 5k and l), but not *FLS1*, *FLS2*, and *FLS3* genes. It is clear that the expression of flavonol synthase genes in Koshu grape skin confers unique characteristics to Koshu grape, enabling the accumulation of large amounts of flavonols in the skin. Flavonoid compounds affect the bitterness of wine (Koyama et al., 2007), suggesting that the astringency of Koshu wines may be also attributed to flavonoid content in Koshu grape skin as well as hydroxycinnamate. Flavan-3-ols are also important for the formation of proanthocyanidins (tannins) and contribute to the taste of red wines. Black-colored Merlot grape skin has high proanthocyanidin content compared to Koshu grape skin (Table 1). Further analysis has demonstrated that Merlot grape skin accumulates much larger amounts of catechin, epicatechin, epicatechin gallate, gallocatechin, and epigallocatechin than Koshu grape skin (Figure 4c). Transcription analysis of chalcone isomerase (*CHI*), dihydroflavonol 4-reductase (*DFR*), leucoanthocyanidin dioxygenase (*LDOX*), and *UFGT* also supports the high accumulation of monomeric and dimeric flavan-3-ols in Merlot grape skin (Figures 5e, f, g, and h).

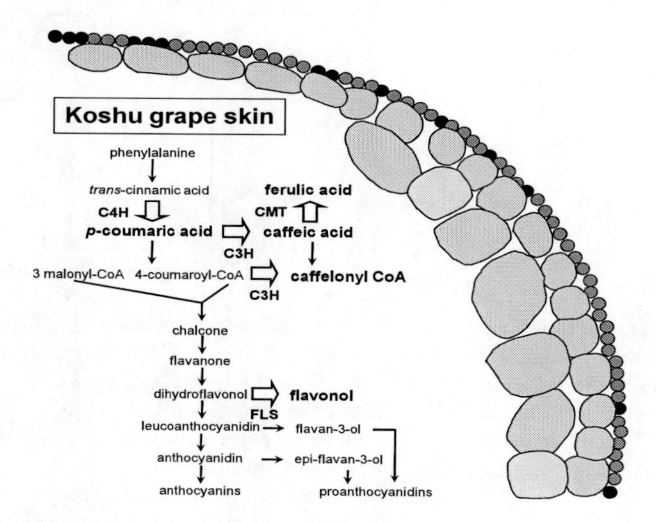

Figure 6. Hypothetical model of total phenolic composition in Koshu grape skin. Large arrows with enzymes in boldface indicate high enzymatic activities in the biosynthesis of phenolic compounds in Koshu grape skin. Compounds in boldface are phenolic compounds that accumulated in Koshu grape skin, as indicated in Table 1 and Figures 4 and 5. Figure adapted from Kobayashi et al. (2011a).

From the above studies, we propose a hypothetical model of total phenolic composition in Koshu grape skin (Figure 6, Kobayashi et al., 2011a). Coumaric acid, caffeic acid, and flavonol contents are present in Koshu grape skin in much larger amounts than those in the other grape cultivars. Flavonol content in grape skin may be influenced by various viticultural practices, such as vine water status (Kennedy et al., 2002), bunch shading (Downey et al., 2004), and exposure to sunlight (Price et al., 1995; Matus et al., 2009). Therefore, canopy management may improve the amount of flavonols in Koshu grape skin and the taste of Koshu wines. These findings are expected to contribute to the improvement of Koshu wines and may serve as a tool to devise new ideas for the development of unique Koshu wines in the future.

Analysis of Flavor Precursors in Koshu Grape

Aroma is one of the most important factors for wine assessment. Because of their low thresholds, volatile thiols emit a powerful odor and are important contributors to wine aroma (Swiegers et al., 2007; Tominaga et al., 1998a). 3-Mercaptohexan-1-ol (3MH) is responsible for passion fruit and grapefruit aroma, and 3-mercaptohexyl acetate (3MHA) is responsible for box tree aroma. These aroma compounds have been well studied as important contributors to the varietal aroma of Sauvignon Blanc (Tominaga et al., 1998a; 1998b), Colombard, Muscat, Sylvaner, Pinot blanc (Tominaga et al., 2000), Petite arvine (Fretz et al., 2005), and rose wines (Murat et al., 2001b). *S*-(3-hexan-l-ol)-L-cysteine (3MH-*S*-cys) was first identified as the precursor of 3MH in Sauvignon blanc juice (Tominaga et al., 1998c). Thereafter, *S*-(3-hexan-1-ol)-glutathione (3MH-*S*-glut) was identified as the tentative pro-precursor of 3MH-*S*-cys (Peyrot des Gachons et al., 2002). Thus, 3MH-*S*-glut is converted into *S*-3-(hexan-l-ol)-L-cysteinylglycine (3MH-*S*-cysgly) via the enzymatic activity of γ-glutamyl transpeptidase. Then, 3MH-*S*-cysgly is transformed into 3MH-*S*-cys by carboxypeptidase (Dubourdieu and Tominaga, 2009).

Finally, 3MH is produced from 3MH-*S*-cys during alcoholic fermentation through the action of yeast β-lyase (Murat et al., 2001a). During alcoholic fermentation, 3MH is partially converted into 3MHA via the enzymatic activity of alcohol acetyltransferase. On the basis of these results, the biosynthetic pathway of 3MH and 3MHA in wine during alcoholic fermentation from glutathionylated pro-precursor-3MH is proposed, as shown in Figure 7.

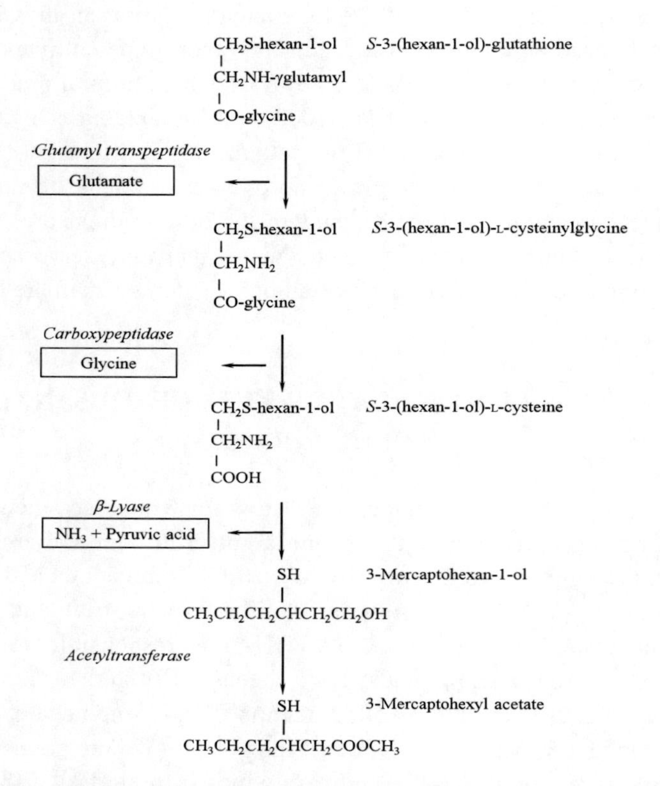

Figure 7. Proposed pathway for the conversion of glutathionylated pro-precursor (3MH-*S*-glut) into the cysteinylated precursor (3MH-*S*-cys) and the production of 3-mercaptohexan-1-ol (3MH) and 3-mercaptohexyl acetate (3MHA) during alcohol fermentation. Figure adapted from Kobayashi et al. (2010).

Moreover, to improve volatile thiol contents in wine, the skin contact method (Maggu et al., 2007), the type of commercial yeast strain used (Dubourdieu et al., 2006), fermentation temperature (Masneuf-Pomarède et al., 2006), and oxygen, phenol, and sulfur dioxide contents (Blanchard et al., 2004), were investigated. As regards varietal aroma in Koshu wines, our previous study has demonstrated the contribution of 3MH and 3MHA in Koshu wines (Kobayashi et al., 2004a). In this chapter, we demonstrate: 1) the localization of 3MH-*S*-glut and 3MH-*S*-cys contents in each Koshu grape tissue and 2) the evolution of 3MH-*S*-glut and 3MH-*S*-cys contents in Koshu grape berries grown in several regions in Japan, during development.

Localization of 3MH-*S*-glut and 3MH-*S*-cys Contents in Koshu Grape Berry

The localization of 3MH-*S*-glut and 3MH-*S*-cys in Koshu grapevine tissues during development is monitored in Figures 8a and b, respectively, using liquid chromatography tandem mass spectrometry (Kobayashi et al., 2010).

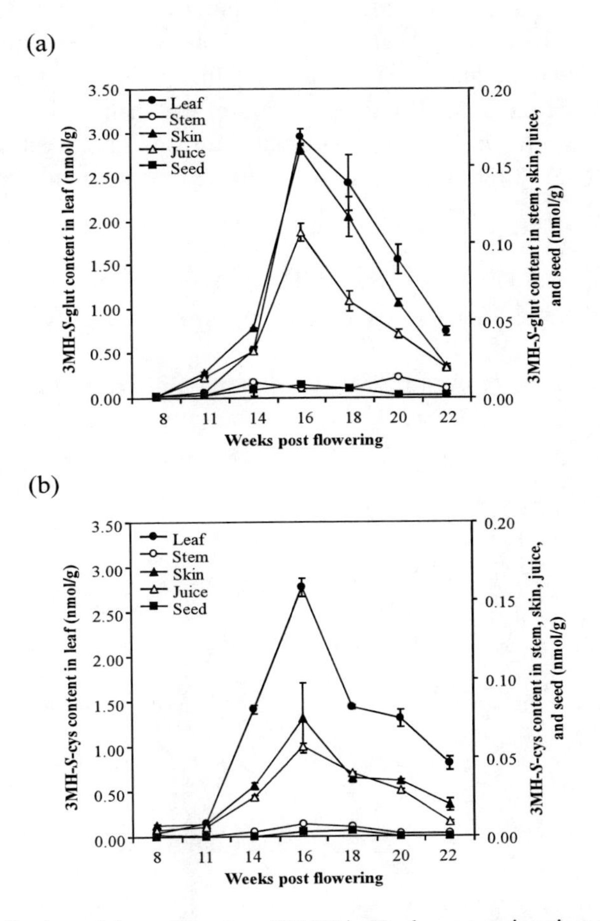

Figure 8. Distribution of the precursors of 3MH in Koshu grapevine tissues during development. 3MH-*S*-glut (a) and 3MH-*S*-cys (b). The Koshu grapevine examined was cultivated in Kofu vineyard. Data are shown as means ± standard deviations from triplicate analysis using three independent tissues. Data adapted from Kobayashi et al. (2010).

Whereas 3MH-*S*-glut and 3MH-*S*-cys were hardly detected in stem and seed throughout the growth period, they were found in leaf, skin, and juice from 11 weeks post flowering (wpf). Unexpectedly, grape leaf has the highest 3MH-*S*-glut and 3MH-*S*-cys contents among the tissues tested, which are approximately 20-fold of those of grape skin and juice throughout the growing period (Figure 8).

Wirth et al. (2001) have reported the presence of various glycoconjugate flavor precursors in grape leaf, such as monoterpenes, norisoprenoids, and shikimic acid related compounds. Large amounts of flavor precursors-3MH also accumulated in grape leaf. However, the reason for the accumulation remains to be determined. The time courses of the changes in 3MH-*S*-glut and 3MH-*S*-cys contents in leaf, skin, and juice are similar. It seems that 3MH-*S*-glut and 3MH-*S*-cys production is synchronized in leaves and berries.

Evaluation of 3MH-*S*-glut and 3MH-*S*-cys Contents in Koshu Grape Berries Grown in Several Regions in Japan

To evaluate the effects of growing region on 3MH-*S*-glut and 3MH-*S*-cys contents in Koshu grape berry, three Koshu grape vineyards having similar soil composition ratios and located in Yamanashi Prefecture, Japan were selected (Figure 9).

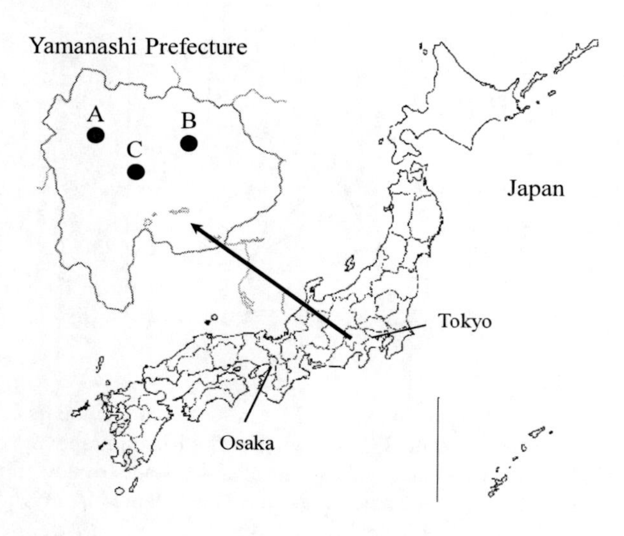

Figure 9. Three test vineyards in Yamanashi Prefecture, Japan.

Table 2. Elevation and soil composition ratio of each vineyard. Data adapted from Kobayashi et al. (2010)

Vineyard	Region	Elevation (m)	Soil composition ratio (%)		
			Sand [a]	Silt [a]	Clay [a]
A	Nirasaki	540	39.9	29.0	31.1
B	Katsunuma	396	57.4	21.6	21.0
C	Kofu	273	43.8	25.3	31.0

[a] Particle size of sand, silt, and clay: 2>0.02, 0.02>0.002, and <0.002 mm, respectively

Those three regions (Kofu, Katsunuma, and Nirasaki) are the main production regions of Koshu grapes cultivated in Japan. Kofu vineyard has the lowest elevation (273 m), whereas Nirasaki vineyard has the highest elevation (540 m). Katsunuma vineyard has moderate elevation (396 m, Table 2). Total soluble solids (TSS, Figure 10a) and titratable acidity (TA, Figure 10b) values suggest that Koshu grape berry grew normally during the growing season. The maturity of grape berry grown in Nirasaki vineyard was delayed by a couple of weeks compared with that grown in Kofu vineyard. Koshu grape berry matured the earliest in Kofu vineyard, which has the lowest elevation among the vineyards tested, whereas maturation was delayed in Nirasaki vineyard, which has the highest elevation. Thus, Koshu grape characteristics are affected by vineyard elevation. Both 3MH-*S*-glut and 3MH-*S*-cys contents in leaf, skin, and juice increased from 8 to 16 wpf. 3MH-*S*-glut and 3MH-*S*-cys contents in grape berries from all the three vineyards peaked between 16 and 18 wpf (Figures 10c and d). 3MH-*S*-glut and 3MH-*S*-cys contents in Koshu grape berries grown in Kofu vineyard peaked the earliest among the three vineyards.

In Katsunuma vineyard, 3MH-*S*-glut and 3MH-*S*-cys contents peaked one week later than those in Kofu vineyard. 3MH-*S*-glut and 3MH-*S*-cys contents in grape berries grown in Nirasaki vineyard were the last to peak (Figures 10c and d). Thus, the accumulation of 3MH-*S*-glut and 3MH-*S*-cys in grape berry shows a similar tendency to grape maturity speed. These results suggest that 3MH-*S*-glut and 3MH-*S*-cys contents may be related to grape berry maturity. Thus, to make Koshu wine having high 3MH and 3MHA contents, grape berries should be harvested at the growing stage in which large amounts of 3MH-*S*-glut and 3MH-*S*-cys are present. To cultivate Koshu grape berry having high 3MH-*S*-glut and 3MH-*S*-cys contents, the relationship between 3MH-*S*-glut and 3MH-*S*-cys contents in grape berry and environmental conditions, such as solar radiation, air temperature, and rainfall, should be determined by further investigations. Sufficient knowledge of the environmental parameters of each vineyard may help determine the appropriate harvest time of grape berries suitable for making aromatic Koshu wine.

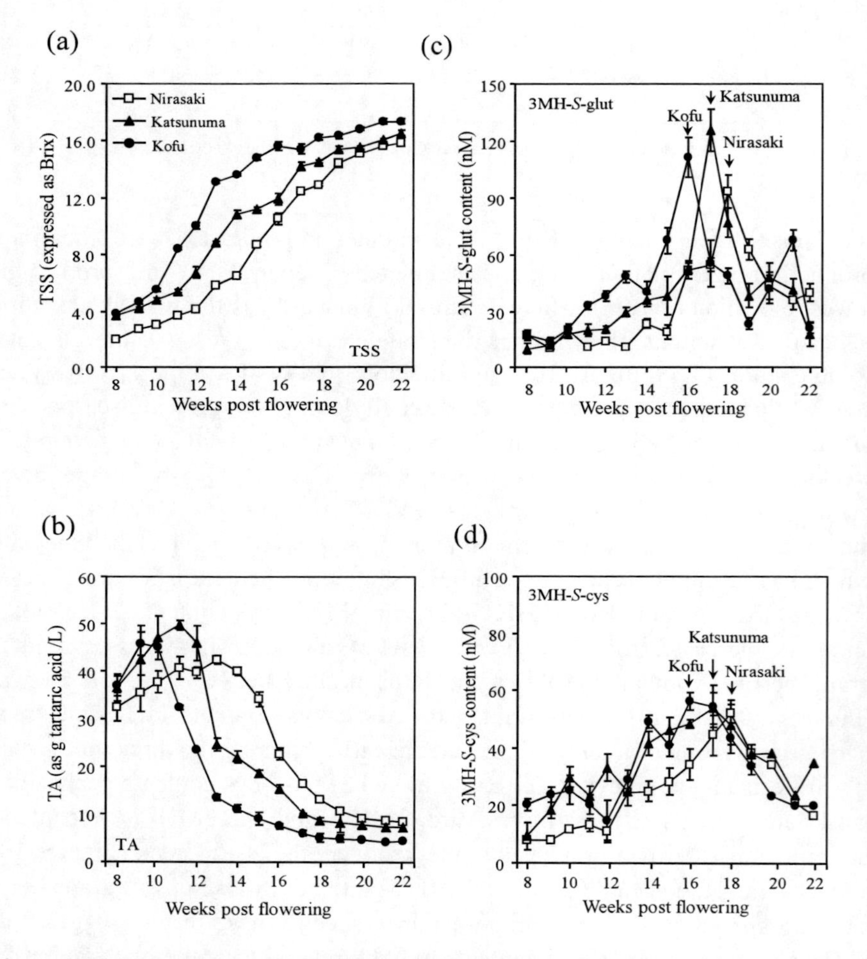

Figure 10. Characteristics of Koshu grape juice collected from the three vineyards during development. (a) TSS, (b) TA, (c) 3MH-*S*-glut content, and (d) 3MH-*S*-cys content. Kofu, ●; Katsunuma, ▲; and Nirasaki, □. Arrows indicate peaks of 3MH-*S*-glut and 3MH-*S*-cys contents for each vineyard. Data are obtained from duplicate analysis of the same juice pressed from 100 to 150 berries. Data adapted from Kobayashi et al. (2010).

Moreover, if leaf plays an important role in the synthesis of 3MH-*S*-glut and 3MH-*S*-cys and the accumulation of these compounds in grape berry, such viticultural practices as leaf removal and/or control of leaf number may influence flavor precursor content in grape berry. These findings may help improve viticultural practices, the final goal of which is to harvest grape berries having high contents of flavor precursors that are responsible for enhancing the varietal aroma of Koshu wine (Kobayashi et al., 2010).

Sarrazin et al. (2007) have reported that 3MH contents in wines made from botrytized grapes infected by *Botrytis cinerea* are much higher than those in wines made from healthy grapes. Thibon et al. (2009) have demonstrated that 3MH-*S*-cys content is much higher in botrytized grapes than in healthy grapes.

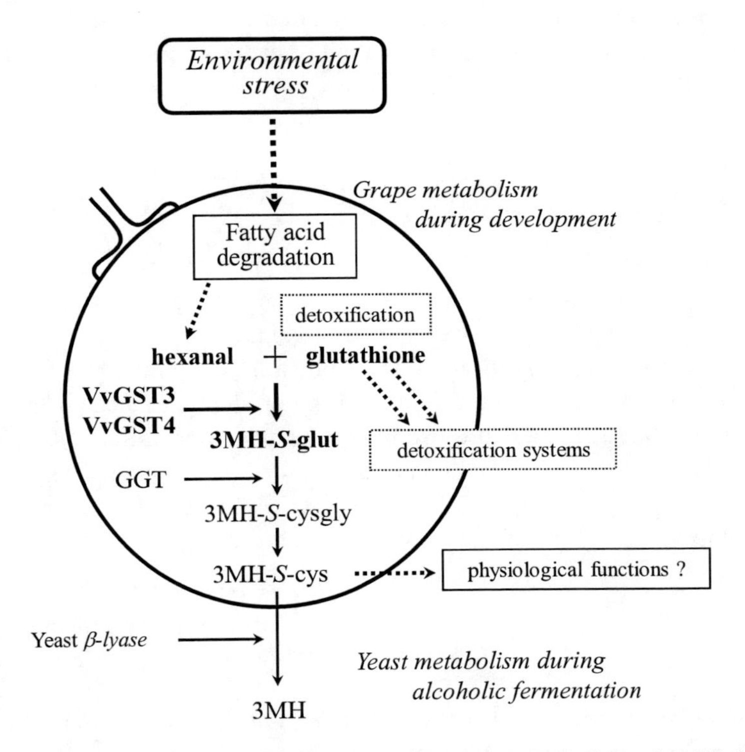

Figure 11. Hypothetical pathway for the biosynthesis of 3MH-*S*-glut and 3MH-*S*-cys in grapevine exposed to environmental stress conditions. Figure adapted from Kobayashi et al. (2011b).

In addition, Thibon et al. (2011) have clarified that *B. cinerea* stimulates the biogenesis of flavor precursors-3MH, including 3MH-*S*-glut, in grapevine cells. These findings suggest that 3MH-*S*-cys biosynthesis in grape berry may be promoted by pathogen attack and, as a result, a high content of 3MH may be present in wines made from botrytized grapes. Moreover, Davoine et al. (2006) have reported that glutathionylated conjugates accumulate in tobacco leaves in response to reactive electrophilic species. Kobayashi et al. (2011b) have demonstrated that environmental stress, such as ultraviolet irradiation, water deficit, cold shock, and biological stimulation, enhances the biosynthesis of 3MH-*S*-glut and 3MH-*S*-cys in grape, and have proposed a hypothetical pathway for the synthesis of 3MH-*S*-glut and 3MH-*S*-cys exposed to environmental stress conditions (Figure 11). The enzymatic mediation of flavor precursors-3MH production by glutathione S-transferase and γ-glutamyl transferase may result in the detoxification of damaged grape cells under stress conditions.

Conclusion

In this review, we demonstrated the physiological and molecular characteristics of Koshu grape. The unique characteristics of Japanese indigenous Koshu grape cultivar were explained by the mechanism of coloration, the transcription analysis of phenolic compounds, distribution of phenolic composition in grape skin, and the production mechanism of flavor precursors. These findings are useful for the selection of suitable cultivation conditions and treatment methods for Koshu grape quality improvement. Further study of the crosstalk between Koshu physiology and enology may yield clues to partially explain 'terroir' of Koshu wines.

References

Ari'Izumi, K., Suzuki, Y., Kato, I., Yagi, Y., Otsuka, K., and Sato, M. (1994) Winemaking from Koshu variety by the sur lie method: Change in the content of nitrogen compounds *American Journal of Enology and Viticulture*, 45, 312-318.

Bogs, J., Ebadi, A., McDavid, D., and Robinson, S. P. (2006) Identification of the flavonoid hydroxylases from grapevine and their regulation during fruit development. *Plant Physiology*, 140, 279-291.

Boss, P. K., Davies, C., and Robinson, S. P. (1996) Analysis of the expression of anthocyanin pathway genes in developing Vitis vinifera L. cv Shiraz grape berries and the implications for pathway regulation. *Plant Physiology*, 111, 1059-1066.

Baranac, J. M., Petranovic, N. A., and Dimitric-Markovic, J. M. (1997) Spectrophotometric study of anthocyan copigmentation reactions. 2. Malvin and the nonglycosidized flavone quercetin. *Journal of Agricultural and Food Chemistry*, 45, 1694-1697.

Blanchard, L., Darriet, P., and Dubourdieu, D. (2004) Reactivity of 3-mercaptohexanol in red wine: impact of oxygen, phenolic fractions, and sulfur dioxide. *American Journal of Enology and Viticulture*, 55, 115 – 120.

Castellarin, S. D., and Gaspero, D. G. (2007) Transcriptional control of anthocyanin biosynthetic genes in extreme phenotypes for berry pigmentation of naturally occurring grapevines. *BMC Plant Biology*, 7, 46.

Castellarin, S. D., Gaspero, G. D., Marconi, R., Nonis, A., Peterlunger, E., Paillard, S., Adam-Blondon, A. F., and Testolin, R. (2006) Colour variation in red grapevines (*Vitis vinifera* L.): genomic organisation, expression of flavonoid 3'-hydroxylase, flavonoid 3',5'-hydroxylase genes and related metabolite profiling of red cyanidin-/blue delphinidin-based anthocyanins in berry skin. *BMC Genomics*, 7, 12.

Castellarin, S. D., Matthews, M. A., Gaspero, G. D., and Gambetta, G. A. (2007) Water deficits accelerate ripening and induce changes in gene expression regulating flavonoid biosynthesis in grape berry. *Planta*, 227, 101-112.

Chatonnet, P., Dubourdieu, D., Noel, J., and Lavigne, V. (1993) Synthesis of volatile phenols by *Saccharomyces cerevisiae* in wines. *Journal of the Science of Food and Agriculture*, 62, 191-202.

Cheynier, V., and Rigaud, J. (1986) HPLC separation and characterization of flavonols in the skin of *Vitis vinifera* var. Cinsault. *American Journal of Enology and Viticulture*, 37, 248-252.

Cohen, S. D., Tarara, J. M., and Kennedy, J. A. (2008) Assessing the impact of temperature on grape phenolic metabolism. *Analytica Chimica Acta*, 621, 57-67.

Cortell, J. M., and Kenndy, J. A. (2006) Effect of shading on accumulation of flavonoid compounds in (Vitis vinifera L.) pinot noir fruit and extraction in a model system. *Journal of Agricultural and Food Chemistry*, 54, 8510-8520.

Davoine, C., Falletti, O., Douki, T., Iacazio, G., Ennar, N., Montillet, J. L., and Triantaphylidès, M. C. (2006) Adducts of oxylipin electrophiles to glutathione reflect a 13 specificity of downstream lipoxygenase pathway in the tobacco hypersensitive response. *Plant Physiology*, 140, 1484-1493.

Dubourdieu, D., and Tominaga, T. (2009) Polyfunctional thiol compounds. *In* Wine Chemistry and Biochemistry. M. V. Moreno-Arribas and M. C. Polo (Ed.), pp. 275-293. Springer Science+Business Media, New York, USA.

Dubourdieu, D., Tominaga, T., Masneuf, I., Peyrot des Gachons, C., and Murat, M. L. (2006) The role of yeasts in grape flavor development during fermentation: The example of Sauvignon blanc. *American Journal of Enology and Viticulture*, 57, 81-88.

Downey, M. O., Harvey, J. S., and Robinson, S. P. (2003) Synthesis of flavonols and expression of flavonol synthase genes in the developing grape berries of Shiraz and Chardonnay (*Vitis vinifera* L.). *Australian Journal of Grape and Wine Research*, 9, 110-121.

Downey, M. O., Harvey, J. S., and Robinson, S. P. (2004) The effect of bunch shading on berry development and flavonoid accumulation in Shiraz grapes. *Australian Journal of Grape and Wine Research*, 10, 55-73.

Fretz, C. B., Luisier, J. L., Tominaga, T., and Amadò, R. (2005) 3-Mercaptohexanol: An aroma impact compound of Petite Arvine wine. *American Journal of Enology and Viticulture*, 56, 407-410.

Fujita, A. (2008) Accumulation and biosynthetic gene expression of flavonols and flavan-3-ols in grape berries. Thesis, Environment and Information Science, Yokohama National University, Yokohama.

Furiya, T., Suzuki, S., Sueta, T., and Takayanagi, T. (2009) Molecular characterization of a bud spot of Pinot gris bearing white berries. *American Journal of Enology and Viticulture*, 60, 66-73.

Glories, Y. (1978) Evolution of phenolic compounds during wine aging. *Annales de la Nutrition et de l'Alimentation*, 32, 1163-1169.

Glories, Y. (1988) Anthocyanins and tannins from wine: organoleptic properties. *Progress Clinical and Biological Research*, 280, 123-134.

Goto-Yamamoto, N., Mochioka, R., Bonian, L., Hashizume, K., Umeda, N., and Horiuchi, S. (1998) RFLP and RAPD analysis of wild and cultivated grapes (*Vitis* spp.) *Journal of the Japanese Society for Horticultural Science*, 67, 483-490.

Goto-Yamamoto, N., Mouri, H., Azumi, M., and Edwards, K. J. (2006) Phenetic clustering of grape (Vitis ssp.) by AFLP analysis. *American Journal of Enology and Viticulture*, 57, 105-108.

Jeong, S. T., Goto-Yamamoto, N., Hashizume, K., and Esaka, M. (2006) Expression of the flavonoid 3'-hydroxylase and 3' 5'-hydroxylase genes and flavonoid composition in grape (*Vitis vinifera*). *Plant Science*, 170, 61-69.

Kennedy, J. A., Matthews, M. A., and Watherhouse, L. (2002) Effect of maturity and vine water status on grape skin and wine flavonoids. *American Journal of Enology and Viticulture* 53, 268-274.

Kobayashi, H., Suzuki, S., Tanzawa, F., and Takayanagi, T. (2009) Low expression of flavonoid 3',5'-hydroxylase (*F3',5'H*) associated with cyanidin-based anthocyanins in grape leaf. *American Journal of Enology and Viticulture*, 60, 362-367.

Kobayashi, H., Suzuki, Y., Ajimura, K., Konno, T., Suzuki, S., and Saito, H. (2011a) Characterization of phenolic compounds biosynthesized in pink-colored skin of Japanese indigenous *Vitis vinifera* cv. Koshu grape. *Plant Biotechnology Reports*, 5, 79-88.

Kobayashi, H., Takase, H., Kaneko, K., Tanzawa, F., Takata, R., Suzuki, S., and Konno, T. (2010) Analysis of *S*-3-(hexan-1-ol)-glutathione and *S*-3-(hexan-1-ol)-L-cysteine in *Vitis vinifera* L. cv. Koshu for aromatic wines. *American Journal of Enology and Viticulture*, 61, 176-185.

Kobayashi, H., Takase, H., Suzuki, Y., Tanzawa, F., Takata, R., Fujita, K., Kohno, M., Mochizuki, M., Suzuki, S., and Konno, T. (2011b) Environmental stress enhances biosynthesis of flavor precursors, *S*-3-(hexan-1-ol)-glutathione and *S*-3-(hexan-1-ol)-L-cysteine, in grapevine through glutathione S-transferase activation. *Journal of Experimental Botany*, 62, 1325-1336.

Kobayashi, H., Tominaga, T., Katsuno, Y., Anzo, M., Ajimura, K., Saito, H., Suzuki, Y., Dubourdieu, D., and Konno, K. (2007) Effect of improvement factor for β-damascenone content in Koshu wine and application to wine making. *Journal of ASEV Japan*, 18, 22-27.

Kobayashi, H., Tominaga, T., Katsuno, Y., Anzo, M., Ajimura, K., Suzuki, Y., Dubourdieu, D., and Okubo, T. (2006) Influence of volatile phenol concentration in Koshu wine quality. *Journal of ASEV Japan*, 17, 75-80.

Kobayashi, H., Tominaga, T., Ueno, N., Ajimura, K., Aruga, Y., Dubourdieu, D., and Okubo, T. (2004a) Key odorous compounds in Koshu wine. *Journal of ASEV Japan*, 15, 109-110.

Kobayashi, S., Goto-Yamamoto, N., and Hirochika, H. (2004b) Retrotransposon-induced mutations in grape skin color. *Science*, 304, 982.

Kobayashi, S., Goto-Yamamoto, N., and Hirochika, H. (2005) Association of *VvmybA1* gene expression with anthocyanin production in grape (*Vitis vinifera*) skin-color mutants. *Journal of the Japanese Society for Horticultural Science*, 74, 196-203.

Kobayashi, S., Ishimura, M., Ding, C. K., Yakushiji, H., and Goto, N. (2001) Comparison of UDP-glucose:flavonoid 3-O-glucosyltransferase (UFGT) gene sequences between white grapes (Vitis vinifera) and their sports with red skin. *Plant Science*, 160, 543-550.

Kobayashi, S., Ishimura, M., Hiraoka, K., and Honda, C. (2002) Myb-related genes of the Kyoho grape (*Vitis labruscana*) regulate anthocyanin biosynthesis. *Planta*, 215, 924-933.

Koyama, K., Goto-Yamamoto, N., and Hashizume, K. (2007) Influence of maceration temperature in red wine vinification on extraction of phenolics from berry skins and seeds of grape (*Vitis vinifera*). *Bioscience, Biotechnology, and Biochemistry*, 71, 958-965.

Masneuf-Pomarède, I., Mansour, C., Marat, M. L., Tominaga, T., and Dubourdieu, D. (2006) Influence of fermentation temperature on volatile thiol concentrations in Sauvignon blanc wine. *International Journal of Food Microbiology*, 108, 385-390.

Maggu, M., Winz, R., Kilmartin, P. A., Trought, M. C., and Nicolau. L. (2007) Effect of skin contact and pressure on the composition of Sauvignon blanc must. *Journal of Agricultural and Food Chemistry*, 55, 10281-10288.

Matus, J. T., Loyola, R., Vega, A., Pena-Neira, A., Bordeu, E., Arce-Johnson, P., and Alcalde, J. A (2009) Post-veraison sunlight exposure induces MYB-mediated transcriptional regulation of anthocyanin and flavonol synthesis in berry skins of *Vitis vinifera*. *Journal of Experimental Botany*, 60, 853-867.

Mori, K., Goto-Yamamoto, N., Kitayama, M., and Hashizume, K. (2007) Loss of anthocyanins in red-wine grape under high temperature. *Journal of Experimental Botany*, 58, 1935-1945.

Murat, M. L., Masneuf, I., Darriet, P., Lavigne, V., Tominaga, T., and Dubourdieu. D. (2001a) Effect of *Saccharomyces cerevisiae* yeast strains on the liberation of volatile thiols in Sauvignon blanc wine. *Am. J. Enol. Vitic.* 52, 136-139.

Murat, M. L., Tominaga, T., and Dubourdieu, D. (2001b) Assessing the aromatic potential of Cabernet Sauvignon and Merlot musts used to produce rose wine by assaying the cysteinylated precursor of 3-

mercaptohexan-1-ol. *Journal of Agricultural and Food Chemistry*, 49, 5412-5417.

Nyman, N. A., and Kumpulainen, J. T. (2001) Determination of anthocyanidins in berries and red wine by high-performance liquid chromatography. *Journal of Agricultural and Food Chemistry*, 49, 4183-4187.

Okamura, S., Watanabe, M. (1981) Determination of phenolic cinnamates in white wine and their effect on wine quality. *Agricultural and Biological Chemistry*, 45, 2063-2070.

Peyrot des Gachons, C., Tominaga, T., and Dubourdieu, D. (2002) Sulfur aroma precursor present in *S*-glutathione conjugate form: Identification of *S*-3-(hexan-1-ol)-glutathione in must from *Vitis vinifera* L. cv. Sauvignon blanc. *Journal of Agricultural and Food Chemistry*, 50, 4076-4079.

Price, S. F., Breen, P. J., Valladao, M., and Watson, T. (1995) Cluster sun exposure and quercetin in Pinot noir grapes and wine. *American Journal of Enology and Viticulture*, 46, 187-194.

Sarrazin, E., Shinkaruk, S., Tominaga, T., Bennetau, B., Frrot, E., and Dubourdieu, D. (2007) Odorous impact of volatile thiols on the aroma of young botrytised sweet wines: Identification and quantification of new sulfanyl alcohols. *Journal of Agricultural and Food Chemistry*, 55, 1437-1444.

Swiegers, J. H., Captone, D. L., Pardon, K. H., Elsey, G. M., Sefton, M. A., Francis, I. L., and Pretorius, I. S. (2007) Engineering volatile thiol release in *Saccharomyces cerevisiae* for improved wine aroma. *Yeast*, 24, 561-574.

Thibon, C., Dubourdieu, D., Darriet, P., and Tominaga, T. (2009) Impact of noble rot on the aroma precursor of 3-sulfanylhexanol content in Vitis vinifera L. cv Sauvignon blanc and Semillon grape juice. *Food Chemistry*, 114, 1359-1364.

Thibon, C., Cluzet, S., Mérillon, J. M., Darriet, P., and Dubourdieu, D. (2011) 3-Sulfanylhexanol precursor biogenesis in grapevine cells : The stimulating effect of *Botrytis cinerea*. *Journal of Agricultural and Food Chemistry*, 59, 1344-1351.

Tominaga, T., Baltenweck-Guyot, R., Peyrot des Gachons, C., and Dubourdieu, D. (2000) Contribution of volatile thiols to the aromas of white wines made from several *Vitis vinifera* grape varieties. *American Journal of Enology and Viticulture*, 51, 178-181.

Tominaga, T., Furrer, A., Henry, R., and Dubourdieu, D. (1998a) Identification of new volatile thiols in the aroma of *Vitis vinifera* L. var. Sauvignon blanc wines. *Flavour Fragrance Journal*, 13, 159-162.

Tominaga, T., Murat, M. L., and Dubourdieu, D. (1998b) Development of a method for analyzing the volatile thiols involved in the characteristic aroma of wines made from *Vitis vinifera* L. cv. Sauvignon blanc. *Journal of Agricultural and Food Chemistry*, 46, 1044-1048.

Tominaga, T., Peyrot des Gachons, C., and Dubourdieu, D. (1998c) A new type of flavor precursors in *Vitis vinifera* L. cv. Sauvignon Blanc: S-cysteine conjugates. *Journal of Agricultural and Food Chemistry*, 46, 52151-5219.

Walker, A. R., Lee, E., Bogs, J., McDavid, D. A. J., Thomas, M. R., and Robinson, S. P. (2007) White grapes arose through the mutation of two similar and adjacent regulatory genes. *The Plant Journal*, 49, 772-785.

Wirth, J., Guo, W., Baumes, R., and Günata, Z. (2001) Volatile compounds released by enzymatic hydrolysis of glycoconjugates of leaves and grape berries from *Vitis vinifera* Muscat of Alexandria and Shiraz cultivars. *Journal of Agricultural and Food Chemistry*, 49, 2917-2923.

Yokotsuka, K. (1990) Effect of press design and pressing pressures on grape juice components. *Journal of Fermentation and Bioengineering,* 70, 15-21.

Yokotsuka, K. (1995) Wine quality and phenolic compounds. *Nippon Shokuhin Kagaku Kogaku Kaishi*, 42, 288-297.

Yokotsuka, K., Matsunaga, M., and Singleton, V. L. (1994) Comparison of composition of Koshu white wines fermented in oak barrels and plastic tanks. *American Journal of Enology and Viticulture*, 45, 11-16.

Yokotsuka, K., Ueno, N., and Singleton, V. L. (2005) Removal of red blush and bitterness from white wine by partial hyperoxidation of juice from the pink-skinned Koshu variety. *Journal of Wine Research*, 16, 233-248.

In: Grapes ISBN 978-1-61470-950-3
Editors: R. P. Murphy et al., pp. 133-153 ©2012 Nova Science Publishers, Inc.

Chapter 5

PROHEXADIONE-CALCIUM AS A REGULATOR OF VINE GROWTH: EFFECT ON THE PHYSICAL-CHEMICAL CHARACTERISTICS OF WINE GRAPES

***A. Gonzalo-Diago*[1,2]*, J. M. Avizcuri*[1,2]*, N. Ortigosa*[1]*, M. Dizy*[2,3]*, M. T. Martínez-Soria*[1]*, J. Sanz-Asensio*[1]*, J. F. Echavarri-Granado*[1] *and P. Fernández-Zurbano*[1,2]**

[1]Department of chemistry, University of La Rioja, Madre de Dios 51, 26006 Logroño, La Rioja, Spain
[2]Wine and Vine Research Institute. (UR-CSIC-GR) Madre de Dios 51, 26006 Logroño, La Rioja, Spain
[3]Department of Agriculture and Food, University of La Rioja, Madre de Dios 51, 26006 Logroño, La Rioja, Spain

ABSTRACT

The production of premium quality grapes requires the control of several and multiple parameters in the vineyard. Therefore, management of production yield is becoming increasingly more important for a quality vitiviniculture. At present, it seems undeniable that high crop yields give lower quality grapes than those obtained from low crop yields. Nowadays, manual cluster thinning is a very widely-used technique to reduce production in vigorous vines or in vines that have been subject to favourable weather conditions. An alternative method for controlling production is the use of plant growth regulators, which has provided successful results in other fruits.

Prohexadione-calcium (Pro-Ca) is a gibberellin biosynthesis inhibitor that produces a decrease in berry size. This leads to an increase in surface-to-volume ratio, theoretically increasing the proportion of phenolic and aroma compounds, very important for grape quality. This product is easily applied by spraying and constitutes no apparent risk to consumers or the environment, because it is reported to be absorbed completely within eight hours and degraded by plants with a half-life of a few weeks and in soil with a half-life of less than one week, without producing toxic metabolites.

The aim of this study was to examine the effect of Pro-Ca application on the physicochemical characteristics of Tempranillo and Grenache grape varieties. Research was carried out in two consecutive harvests 2007 and 2008. The effect of Pro-Ca on grapes was compared with grapes cultivated using manual cluster thinning and control vines. Application of Pro-Ca in both varieties shows a decrease in berry size and weight. The grapes obtained from treated vines, both manual cluster thinning and Pro-Ca, presented a higher concentration of phenolic compounds and an enhancement of colour than those from the control. Evolution of alcoholic fermentation of must from grapes treated with Pro-Ca was correct and similar to manual cluster thinning and untreated grapes.

Keywords: Wine grapes, Prohexadione calcium, Cluster thinning, Growth regulator.

1. INTRODUCTION

The production of premium quality grapes requires the control of several and multiple parameters in the vineyard. Therefore management of production yield is becoming increasingly more important for a quality vitiviniculture (Petrie and Clingeleffer, 2006). So far, manual cluster thinning is a widely used technique to reduce production in vigorous vines, producing an increasing in Colour Intensity (CI), Total Phenolic Index (TPI) and anthocyanins (Garcia-Escudero *et al.*, 2004). Although effective, manual cluster thinning is a very expensive operation because of the large labour requirements (Martinez de Toda *et al.*, 2003).

An alternative method for controlling production is the application of plant growth regulators, which has provided successful results in other fruits.

Plant growth regulators are an important chemical tool since they allow the control of several physiological parameters involved in vegetative and reproductive developments (Sang-Mo Kang *et al.*, 2010). Among them, prohexadione-calcium (Pro-Ca) (calcium 3,5-dioxo-4 propionylcyclohexanecarboxylate, Figure 1) is of particular relevance (Costa *et al.*, 2006).

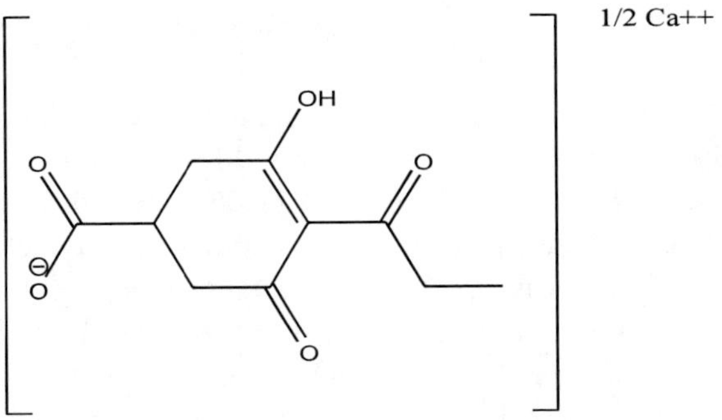

Figure 1. Prohexadione-calcium structure.

Pro-Ca belongs to the group acylcyclohexanediones and it is a structural mimic of 2-oxoglutaric acid, which is the co-substrate of dioxygenases that catalyse late steps of gibberellins (GAs) formation. Pro-Ca blocks particularly 3β-hydroxylation, thereby inhibiting the formation of biologically active GAs from biologically inactive GAs (Rademache, 2000). Interference of Pro-Ca with the metabolism of gibberellins, ethylene and flavonoids is of particular importance due to their practical relevance. Such an interference results in lowering the content of bioactive gibberellins and reducing ethylene formation, which causes reduction in vegetative growth and an early fruit production (Owens and Stover, 1999), and hence enhances the quality and yield (Sang-Mo Kang *et al.*, 2010).

As Pro-Ca is able to alter flavonoid metabolism, novel flavonoids are formed that were previously identified as 3-deoxycatechins in young leaves of apple (Römmelt *et al.*, 1999 and 2003a), pear (Römmelt *et al.*, 2003b), grapevine leaves (Gosch *et al.*, 2003) and in berries (Puhl *et al.*, 2008). More research on the effect of Pro-Ca in the phenolic composition of grapes and if this application affects the evolution of alcoholic fermentation is needed and is the aim of the present study.

Besides, the alteration of flavonoid composition is proposed to be related to enhanced resistance of several crop plants (Römmelt *et al.*, 2003a, b; Gosch *et al.*, 2003, Rühmann *et al.*, 2003, Fischer *et al.*, 2006). Pro-Ca itself does not possess any antimicrobial activity, but its application leads to a reduction of the infection caused by the bacterium *Erwinia amylovora* in apples (Römmelt *et al.*, 1999, Yoder *et al.*, 1999, McGrath *et al.*, 2009) and pears (Costa *et al.*, 2001). Many studies suggest that the phenolic compounds can be involved in plant resistance as antimicrobial compounds (Scalbert *et al.*, 1991, Yamamoto *et al.*, 2000, Del Rio *et al.*, 2003, Parvez *et al.*, 2004) or as mechanical barriers (Weber *et al.*, 1995, Feucht *et al.*, 1998, Schwalb *et al.*, 1998, Feucht and Treutter, 1999).

Pro-Ca is easily applied by spraying and constitutes no apparent risk to the consumer or the environment (Rademacher and Kober, 2003). It also has a low propensity for crop residues (Winkler, 1997). This growth regulator has been patented by Kumiai Chemical Industry Co. and has been registered for growth control of rice (*Oryza sativa* L.) in Japan (Evans *et al.*, 1999). Recently, Pro-Ca has been registered in the United States (under the trade name Apogee), as a replacement for daminozide, for use on apples (Malus x domestica Borkh.) and as Regalis in Europe (Miller and Tworkoski, 2003).

Disegna *et al.* (2003) in cv. Tannat observed that the treatment with Pro-Ca produced a reduction in the vegetative growth, grape production and berry weight and size. The wines were also more fruited, with higher alcohol content, higher colour intensity, higher mouth volume and with more complexity, persistence and aromatic intensity. These wines presented higher terpene and norisoprenoid contents with respect to the control. Subsequently, Lo Giudice *et al.* (2004) focused on other varieties, such as cv. Cabernet Sauvignon. They also found a reduction in berry weight following the application of Pro-Ca, which is correlated with an increase in colour intensity, total anthocyanins and total phenols. The effects observed on grape composition were generally positive, but the effect on the quality and the organoleptic characteristics of the final wine are still unknown by these authors. In a previous work performed by our research group in cv. Tempranillo the treatment with Pro-Ca led to a reduction in yield, clusters and berry size. Moreover, better and earlier grape ripening was achieved with a high Brix degree. An increase in TPI, tannins and CI in the wines was observed following the treatment of the grapes. Duo-trio tests used for sensory analysis did not show significant differences at a level of 5 % but the application of Pro-Ca enhanced the typical sensory attributes of Tempranillo wines (Vaquero et al., 2009).

The effect of Pro-Ca application on the phenolic composition of Tempranillo and Grenache varieties is not well known. This study aims to obtain a better knowledge of the effect of Pro-Ca on the phenolic composition of grapes, including another variety, Grenache, which is one of the most widely planted red wine grape varieties in the world.

2. Materials and Methods

2.1. Chemicals

(+)-Catechin was supplied by Extrasynthese (Genay, France). Siringaldazine, polivinylpolipirrolidone and ethanol absolute were purchased from Sigma Aldrich (Steinheim, Germany). Methylcellulose was from Acros Organics (New Jersey, USA). Sodium acetate, sodium metabisulfite, ammonic sulphate and clorhidric acid were obtained from Scharlab (Barcelona, Spain). Deionized water was purified with a mili-Q water system (Millipore, Molsheim, France) prior to use. Reference standard for red wine, with values certified to the classical parameters by the Ministry of Agriculture, Fishing and Food, was obtained from Panreac (Barcelona, Spain).

2.2. Vineyards

The experiments were conducted in two commercial vineyards with Tempranillo and Grenache (both *Vitis vinifera* L.) non-irrigated and situated in two nearby locations of the Rioja Qualified Denomination of Origin (Spain) over two consecutive harvests, 2007 and 2008.

Site 1 was located in Aldeanueva de Ebro (Southeastern Rioja, 42°13′N 1°53′O, 346m). Tempranillo vines were planted using a gobelet training system. Vine spacing was of 2.6 m (row) x 1.2 m (vine). Vines were cane-pruned (twelve buds per vine), grafted onto 110R rootstock (clone 51) and planted in 2000 with an extension of 1 ha.

Site 2 was located in Rincón de Soto (Southeastern Rioja, 42°14′4″N 1°51′3″O / 42.23444, -1.85083, / 42.23444, -1.85083291 m) Grenache vines (twelve buds per vine) were trained and spur-pruned on a wire trellis system for support and the canopy was vertically shoot-positioned. Row and vine spacing was of 2.6 m x 1.2 m. Vines were grafted onto 11O R rootstock (clone 70) and planted in 1998 with an extension of 2.5 ha.

Both of them presented constant conditions of vegetative growth and development under similar conditions of fertilization and irrigation. Shoot density was adjusted to 12 shoots per vine in both cases.

2.3. Field Treatments

Prohexadione-Ca was applied as Regalis®, 10 % Pro-Ca, (BASF, Germany) at a single dose of 3 kg ha^{-1} at pre-blooming (BBCH57) using a HARDI® atomizator (Denmark). Applications were made when there was no rain predicted for a minimum of 24 hours. Treatments were applied to both sides of the grapevine canopy, wetting the entire shoots. Three treatments were trialled: control vines (unthinned), manual cluster thinning vines (30 % in Tempranillo and 50 % in Grenache in both years, except 2008 when manual cluster thinning was not carried out in Tempranillo variety because of the low production of grapes on the vines), and vines treated with prohexadione of calcium.

In both sites, treatments were arranged in a completely randomised design that consisted of six rows for each treatment. Random vines of the two central rows of each treatment were sampled with the aim of avoiding interferences with treatments made in other vines.

2.4. Sample Preparation

Fifty berry samples were collected from each treatment in triplicate immediately after harvest to determine average berry weight. One hundred berry samples were collected to determine size and the solid/liquid ratio. Berry sampling in all experiments consisted of a random selection of clusters, but with a methodical sampling of individual berries along the cluster, top, middle and the bottom of the cluster.

Three replicates of the fifty berry samples from each treatment were frozen and stored at -20 °C for fruit composition analysis.

2.5. Technological Ripeness

For the study of technological ripeness, 100 berries, randomly sampled and in triplicate, both control and treated grapes, were weighed and placed in trays with different diameter holes (18,16,14,12 and 10 mm). Thus, berries passed through from higher diameter trays to lower ones. Berries were manually crushed and filtered. Then solid was separated and weighed and the liquid volume was measured. Conventional oenological parameters (density, °Brix, pH, total acidity and malic acid) were determined with WineScan FT120 (FOSS, Denmark). Amount of laccase was determined using the method described by Dubourdieu *et al.* (1984).

2.6. Phenolic Ripeness

For the study of phenolic ripeness, three replicates of 50 berries, randomly sampled, were prepared according to Iland *et al.* (2004). This methodology is based on the maceration of the berries crushed in a methanolic solution (50 %) during an hour. Then, polyphenols and anthocyanins are evaluated measuring its absorbance at 280 and 520 nm, respectively and in acidic medium. Tannins are measured after precipitation with methylcellulose, quantified against a catechin standard curve and expressed as milligrams of catechin equivalents per gram fresh weight of skin. Anthocyanins are expressed as milligrams of malvidine-3-glucoside per gram fresh weigh of skin. Tristimulus parameters (L*, C* and H*) were calculated to obtain the chromatic coordinates CIELab, L* (lightness), a* (red/green) and b* (yellow/blue), using a standard illuminator D_{65} and a standard observation CIE 1964 based on the method of *OIV (*Ayala, *et al.*, 1997). The visible spectra (380 - 770 nm) data were acquired using quartz cell of 1 mm path length. All analyses were done in triplicate.

2.7. Vinification Process

The grapes were harvested manually in their different production areas on the optimum harvest date in 15 kg plastic boxes and quickly transported to the experimental winery of the University of La Rioja. After crushing and destemming, grapes and musts were transferred to closed 50 L stainless steel tanks to undergo alcoholic fermentation. The musts were treated with sulphur dioxide (5 - 6.5 g/hL) and inoculated with active dry yeast *Saccharomyces cerevisiae* (30 g/hL) (VRB Uvaferm®, Lallemand, Australia) The fermenting musts were punched down twice a day, temperature was maintained below 28°C for 10 days and both temperature and density were measured daily.

Subsequently, wines were transferred to closed 15 L stainless steel tanks to undergo malolactic fermentation induced by inoculation with a commercial preparation of *Oenococcus oeni* (1 g/hL) (MRB® Alpha, Lallemand, Denmark). All the wines were racked, treated with sulphur dioxide (5 g/hL) and bottled.

2.8. Statistical Analysis

All data obtained were assessed by one-way analysis of variance (ANOVA) (InfoStat 2007 edition; Cordoba, Argentina and Excel 2003) to identify significant differences between control and treated grapes. Differences between samples always refer to significant differences with at least $P < 0.05$.

3. Results And Discussion

3.1. Weight and Size of Berries

The effectiveness in yield reduction by manual and chemical thinning in both varieties and seasons is shown in Table 1. The results show that the application of Pro-Ca produced a reduction in the crop yield in Grenache of 27 % and 16 % in 2007 and 2008 respectively and in Tempranillo 13 % in 2007, whereas there was no reduction in 2008. This yield has been calculated from their average berry weight.

In table 1, the size and solid/liquid relation of the grapes under study are shown. In all cases, except for the Tempranillo 2008, the grapes from the strain treated with Pro-Ca showed a weight significantly lower than the control grapes. This agrees with the findings of Lo Giudice *et al.* (2004) and Vaquero *et al.* (2009). Less regular behaviour was shown in the grapes from the strains that had undergone manual thinning, just as in 2007, the grapes from the Tempranillo variety, having undergone this treatment, had a significantly greater weight than the control grapes and those treated with Pro-Ca. In the Grenache variety, in year 2007, the grapes from manual thinning did not show significant differences with the control grapes, while in 2008, the size of the grapes from manual thinning were significantly smaller than the control grapes and significantly larger than the grapes treated with Pro-Ca. Similar results have been obtained regarding the size of the berries (Table 1).

Table 1. Effect of cluster thinning and Pro-Ca treatment on Tempranillo and Grenache berry weight, size and solid/liquid relation at harvest

	Berry weight (g)	*Berry size* (diameter, mm)*					% Solid/liquid (m/v)
		10-12	*12-14*	*14-16*	*16-18*	*18-20*	
TEMPRANILLO							
2007							
Control	1.96 ± 0.02^{b}	8 ± 0^{b}	54 ± 7^{a}	32 ± 7^{b}	8 ± 0^{a}	0 ± 0^{a}	42.2 ± 2.4^{b}
Cluster Tinning	2.13 ± 0.02^{a}	0 ± 1^{c}	44 ± 3^{b}	50 ± 3^{a}	6 ± 1^{a}	0 ± 0^{a}	59.7 ± 4.1^{a}
Pro-Ca	1.71 ± 0.01^{c}	24 ± 2^{a}	57 ± 4^{a}	19 ± 2^{c}	3 ± 1^{b}	0 ± 0^{a}	57.1 ± 4.3^{a}
2008							
Control	$2,27 \pm 0.10^{a}$	0 ± 0^{a}	24 ± 6^{a}	59 ± 5^{a}	16 ± 7^{a}	1 ± 1^{b}	$58,4 \pm 9.5^{a}$
Pro-Ca	$2,31 \pm 0.19^{a}$	0 ± 0^{a}	19 ± 4^{a}	52 ± 8^{a}	25 ± 4^{a}	5 ± 2^{a}	$52,6 \pm 0.6^{a}$
GRENACHE							
2007							
Control	2.36 ± 0.02^{a}	0 ± 0^{a}	31 ± 6^{b}	54 ± 5^{a}	16 ± 4^{a}	0 ± 0^{a}	55.2 ± 1.3^{a}
Cluster Tinning	2.36 ± 0.13^{a}	0 ± 0^{a}	38 ± 7^{b}	51 ± 4^{a}	11 ± 3^{b}	0 ± 0^{a}	53.1 ± 2.1^{ab}
Pro-Ca	1.72 ± 0.05^{b}	3 ± 0^{b}	64 ± 4^{a}	27 ± 5^{b}	6 ± 1^{c}	0 ± 0^{a}	50.0 ± 2.5^{b}
2008							
Control	$1,99 \pm 0.08^{a}$	0 ± 0^{a}	57 ± 2^{c}	38 ± 4^{a}	6 ± 2^{a}	0 ± 0^{a}	73.4 ± 5.1^{a}
Cluster Tinning	$1,86 \pm 0.09^{b}$	0 ± 0^{a}	64 ± 2^{b}	32 ± 3^{a}	4 ± 2^{ab}	0 ± 0^{a}	71.6 ± 5.1^{a}
Pro-Ca	$1,67 \pm 0.06^{c}$	0 ± 0^{a}	74 ± 3^{a}	24 ± 4^{b}	2 ± 1^{b}	0 ± 0^{a}	75.1 ± 4.6^{a}

* % of berries for each diameter; a,b,c Different letters in same column and year means significant differences at 5 % level.

The berries treated with Pro-Ca were significantly smaller than those of the control strain and manual thinning, while more random behaviour was observed in this last treatment.

Concerning the solid/liquid relation, differences were only found in the grapes from 2007, but with different relationships in both varieties of grape studied. Thus, while in the Tempranillo variety the thinning and Pro-Ca treatments showed a better solid/liquid relation than the control grapes in the Grenache variety, only the control grapes and those from the treatment with Pro-Ca, showed significant differences, being in this case the control grapes that had a better solid/liquid relationship.

As previously mentioned (section 2.3) in 2008, and in the case of the Tempranillo variety, manual thinning could not be carried out due to the low grape production. The non-existence of significant differences in the weight, size and solid/liquid relation between the grapes from the control strains and those from the treatment with Pro-Ca could either be because the application of prohexadione-calcium did not affect the development of the fruit, or because the control strains underwent a natural development similar to the application of a production decreasing method.

3.2. Technological Ripeness

In both grape-growing seasons and varieties, laccase values were low, showing no significant differences among samples (Table 2). Vail and Marois (1991) have described a significant correlation between cluster compactness and the susceptibility to bunch rots caused by *Botrytis cinerea.* The application of Pro-Ca provided smaller berries reducing this effect as described by Vaquero *et al.* (2009).

In accordance with the other parameters analysed and shown in table 2, the differences of the applied treatments with the control samples were again greater in 2007 than in 2008, for which no significant differences were found between the treatments. Undoubtedly, the most important difference is that obtained in the sugar content (ºBrix). In both cases and varieties, the grapes from the natural thinning showed the higher sugar content. The advancement of grape ripeness as a result of thinning has been documented in different varieties, such as Cabernet Sauvignon (Petrie and Clingeleffer 2006), Shiraz (Clingeleffer *et al.*, 2002), Tempranillo (Tardáguila *et al.*, 2008) and Grenache (Diago *et al.*, 2010). The values ºBrix, pH, total acidity and malic acid of the grapes in the treatment with Pro-Ca did not show significant enological differences with the control grapes. These results agree with other studies published on this issue, which indicate that the application of Pro-Ca did not cause changes in these parameters. (Lo Giudice *et al.,* 2004, Vaquero *et al.*, 2009).

Table 2. Effect of cluster thinning and Pro-Ca treatment on Tempranillo and Grenache technological ripeness parameters

	° Brix	pH	Titratable acidity[c]	Malic acid (g/l)	Lacasse[d]
TEMPRANILLO					
			2007		
Control	22.8 ± 0.1[b]	3.53 ± 0.08[a]	6.26 ± 1.07[b]	2.43 ± 0.11[a]	0.00 ± 0.00[a]
Cluster Tinning	24.1 ± 0.4[a]	3.41 ± 0.02[b]	6.32 ± 1.29[b]	2.19 ± 0.04[b]	0.18 ± 0.26[a]
Pro-Ca	23.5 ± 0.6[b]	3.38 ± 0.02[b]	6.83 ± 0.05[a]	2.28 ± 0.05[ab]	0.08 ± 0.05[a]
2008					
Control	24.3 ± 0.9[a]	3.40 ± 0.05[a]	4.14 ± 0.18[a]	1.70 ± 0.16[b]	0.00 ± 0.00[a]
Pro-Ca	24.4 ± 1.1[a]	3.34 ± 0.01[a]	4.69 ± 0.37[a]	1.96 ± 0.09[a]	0.18 ± 0.08[a]
GRENACHE					
			2007		
Control	20.60 ± 0.5[b]	3.24 ± 0.02[b]	6.78 ± 0.06[a]	1.73 ± 0.03[a]	3.43 ± 4.11[a]
Cluster Tinning	25.33 ± 0.8[a]	3.29 ± 0.02[a]	6.09 ± 0.24[b]	1.15 ± 0.18[b]	2.25 ± 1.94[a]
Pro-Ca	20.83 ± 1.8[b]	3.26 ± 0.03[ab]	6.54 ± 0.16[a]	1.92 ± 0.23[a]	0.25 ± 0.39[a]
2008					
Control	25.54 ± 0.7[a]	3.26 ± 0.02[a]	5.56 ± 0.20[a]	1.28 ± 0.20[a]	1.71 ± 0.93[a]
Cluster Tinning	25.84 ± 0.3[a]	3.27 ± 0.02[a]	5.40 ± 0.11[a]	1.04 ± 0.08[b]	0.91 ± 0.15[a]
Pro-Ca	25.51 ± 0.7[a]	3.25 ± 0.04[a]	5.68 ± 0.38[a]	0.93 ± 0.05[b]	1.29 ± 0.48[a]

[a,b] Different letters in same column and year means significant differences at 5 % level.
[c] expressed as g tartaric acid per litre.
[d] expressed as nmol lacasse per litre.

3.3. Phenolic Ripeness

As was to be expected the phenolic content was in all cases significantly higher in Tempranillo than in Grenache so the distribution of these compounds in grapes mainly depends on variety (Figure 2, 3 and 4).

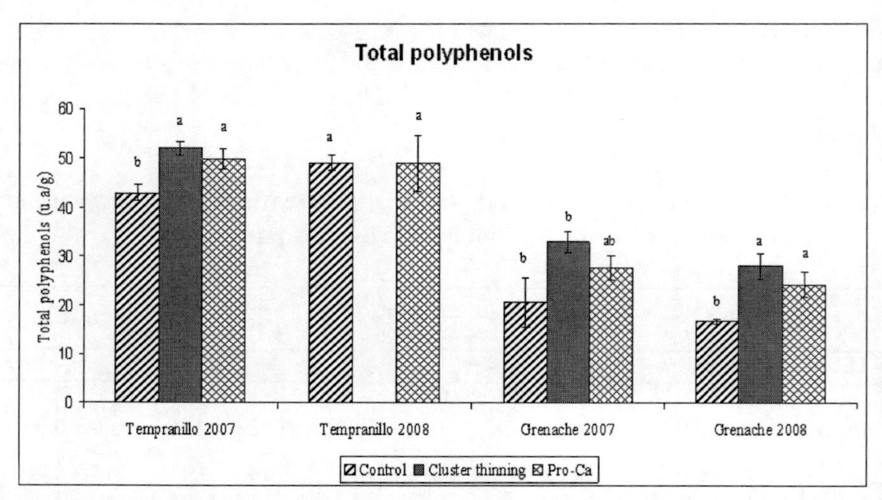

Different letters in the same variety and year means significant differences at 5 % level.

Figure 2. Content of total polyphenols for both varieties, years and treatments.

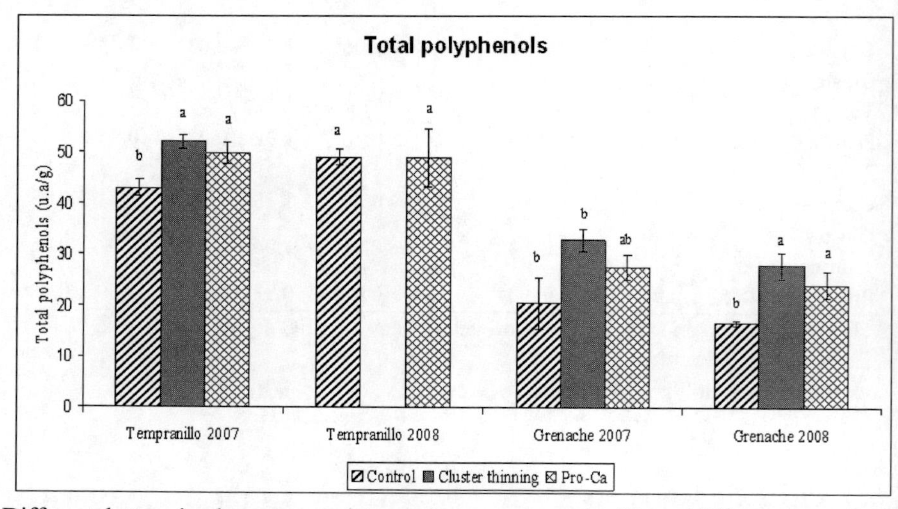

Different letters in the same variety and year means significant differences at 5 % level.

Figure 3. Content of total tannins for both varieties, years and treatments.

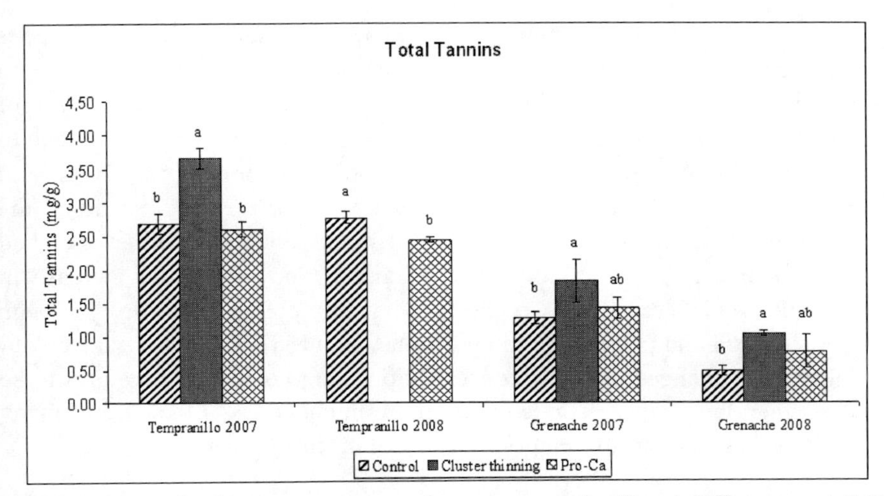

Different letters in the same variety and year means significant differences at 5 % level.

Figure 4. Content of total anthocyanins for both varieties, years and treatments.

Nevertheless, almost all references coincide with the fact that the non-genetic factors such as climate, soil conditions or viticultural practices have an effect on the concentration of polyphenols (Arozarena et *al.*, 2002).

This influence can be observed for each variety in the two years studied, as the polyphenol content was higher in the grapes from 2007 than in those from 2008.

As regards the total polyphenol content (Figure 2), it can be observed that in both varieties the treatments performed to reduce the harvest resulted in grapes with a higher polyphenol content than in the control grapes. Nevertheless, both treatments have a different effect on the anthocyanin and tannin content in the grapes. The Tempranillo grapes from the strains subjected to manual cluster thinning showed a higher tannin content (Figure 3), while those from the treatment with Pro-Ca showed a higher anthocyanin content (Figure 4).

The results of the polyphenolic composition in the Grenache variety were very similar over the two years of study (Figures 2, 3 and 4). The grapes from the strains subjected to manual thinning showed higher tannin and anthocyanin content than the control grapes, while the grapes from the Pro-Ca treatment did not show any significant differences, either in the control grapes or in those from the manually treated strains, showing that the application of Pro-Ca resulted in grapes with a polyphenol content midway between the control grapes and those from the thinning. Both the results found in the two varieties and those in the two years studied, showed that the application of the growth regulator does not cause, in any cases, any decrease in the synthesis of polyphenolic compounds. Similar findings in phenolic content increase due to mechanical thinning were observed in Nebbiolo (Guidoni *et al.,* 2002) and Cabernet Franc (Mazza *et al.*, 1999). Peña-Neira *et al.* (2007) studied the changes in phenolic composition of Syrah grape skins from vines that had undergone manual cluster thinning at veraison. They found an increase in some phenolic derivatives related to wine colour stability through copigmentation and polymerisation reactions (Boulton 2001). Red wine colour stability and enhanced phenolic concentrations are two essential features to be sought when long aging periods in barrel are intended, as is the case of many Spanish wines made from Tempranillo and Grenache grapes.

3.4. Chromatic Parameters

It is well known that colour intensity fell with ageing, with a simultaneous increase in lightness (L*) and decrease of violet tonalities (b*), and a decrease in red hue (a* values). Although in this case, samples studied are musts, this can be a first approximation and a useful tool for predicting what the colour of future wines will be. The study of colour is interesting because both colour and phenolic content are important indicators of red wine quality (Iland 1987).

In the analysis of colour made by the study of CIELab parameters, the only year in which there were not significant differences in colour was in 2007 for Grenache variety (Table 3), but it is necessary to say that this variety tends to oxidise easily, and, in that year, samples suffered a rapid oxidation that is reflected in the extremely high values that show the coordinate b* in the three cases.

In 2007, in the case of Tempranillo, and in 2008, for both varieties, the musts following cluster thinning and the vines treated with Pro-Ca showed a minor value of L* and more red and blue hue. This outcome might be explained by the significant yield drop caused by the application of both treatments. Likewise, other studies have shown that the colour of red wine is usually reduced when the yield is high (Bravdo *et al.*, 1984), whereas low yield levels can increase anthocyanins and total phenolic compounds of fruit and wine (Mazza *et al.*, 1999, Guidoni *et al.*, 2002). Several authors have also reported improvements in wine colour density as a result of manual and mechanical cluster thinning (Clingeleffer *et al.*, 2002, Petrie and Clingeleffer 2006).

Table 3. Chromatic parameters in Tempranillo and Grenache grapes

	L*	a*	b*	IC	Hue
TEMPRANILLO					
			2007		
Control	58.1 ± 0.1[a]	33.8 ± 0.9[c]	-5.9 ± 0.7[b]	3.16 ± 0.03[a]	0.74 ± 0.02[a]
Cluster Tinning	54.1 ± 1.2[b]	36.2 ± 1.1[b]	-4.37 ± 1.1[a]	3.61 ± 0.15[a]	0.76 ± 0.00[a]
Pro-Ca	48.4 ± 1.2[c]	44.0 ± 0.7[a]	-8.7 ± 1.8[c]	3.83 ± 0.65[a]	0.67 ± 0.04[b]
2008					
Control	96.0 ± 0.9[a]	9.07 ± 0.7[a]	-0.56 ± 0.2[a]	1.18 ± 0.45[b]	0.79 ± 0.07[a]
Pro-Ca	86.9 ± 2.0[b]	10.55 ± 1.7[a]	-2.89 ± 0.7[b]	4.04 ± 0.64[a]	0.77 ± 0.02[a]
GRENACHE					
			2007		
Control	76.3 ± 2.9[a]	21.1 ± 1.2[a]	3.1 ± 1.6[a]	1.83 ± 0.28[a]	0.96 ± 0.07[a]
Cluster Tinning	72.4 ± 4.2[a]	21.6 ± 4.1[a]	6.1 ± 2.9[a]	2.15 ± 0.31[a]	1.04 ± 0.09[a]
Pro-Ca	75.3 ± 2.8[a]	18.8 ± 2.5[a]	5.2 ± 2.6[a]	1.95 ± 0.20[a]	1.11± 0.09[a]
2008					
Control	94.9 ± 0.3[a]	3.56 ± 0.4[c]	1.23 ± 0.2[a]	1.73 ± 0.08[b]	1.08 ± 0.06[a]
Cluster Tinning	87.8 ± 0.8[c]	6.14 ± 0.1[a]	1.98 ± 0.6[a]	4.86 ± 0.90[a]	1.09 ± 0.04[a]
Pro-Ca	92.6 ± 0.1[b]	4.98 ± 0.0[b]	0.22 ± 0.1[b]	2.81 ± 0.62[b]	0.95 ± 0.01[b]

[a,b] Different letters in same column and year means significant differences at 5 % level.

With regard to colour intensity and hue, in 2007 we did not observe differences but in 2008 the grapes treated showed higher colour intensity and lower hue than control ones. It is important to note that colour intensity and hue are according to CIELab parameters.

3.5. Evolution of Alcoholic Fermentation

With regard to the fermentation process, the figure 5 shows the evolution of sugars in 2007 for both varieties. The differences between unthinned and thinned vines (manually and chemically) were only observed for the Tempranillo variety. No significant differences in the evolution of fermentation were identified for both treatments and varieties in 2008 (data not shown). Fermentation behaviour was correct in all cases being the rate of change in density higher in musts obtained from treated grapes. No stuck or sluggish fermentation was observed and the remaining residual sugar was lower than 5 g/L.

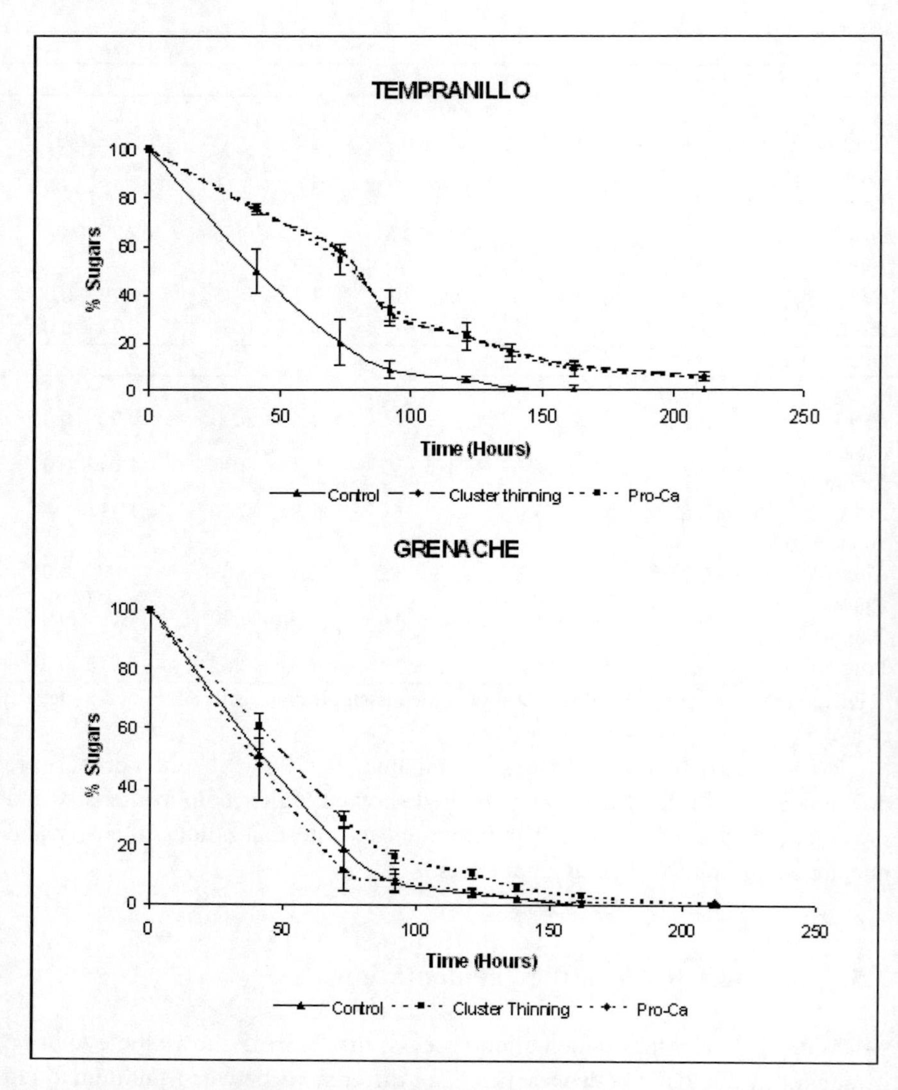

Figure 5. Evolution of the % of sugar during alcoholic fermentation of musts from the studied treatments in 2007.

CONCLUSION

In conclusion, it can be said that the pre-blooming application of Pro-Ca produced a reduction in yield, clusters and berry size in cv. Tempranillo and Grenache. Technological parameters, in general, were not modified. Alcoholic fermentation was not affected either, this took place without problems in all cases.

The total content of polyphenols did not turn out to be diminished by the application of the growth regulator. On the contrary, the content of total polyphenols in both varieties was higher in must produced from vines thinned, whether manually or chemically, than the control vines. New research will be necessary in order to evaluate the different impact that the addition of the growth regulator has had on the anthocyanic composition of the grapes, which was similar to the concentration of the manually thinned samples, while the tannin content was lower.

Finally, the CIELab parameters concluded that cluster thinning and the application of Pro-Ca made the musts more intense and darker (lower L*) and with higher red and violet tonality. Only in 2007, in musts of Grenache, there were no differences between them, due to samples suffering a rapid oxidation that year.

REFERENCES

Arozarena, I., Ayestarán, B., Cantalejo, M.J., Navarro, M., Vera, M., Abril, I. and Casp, A. (2002). Anthocyanin composition of Tempranillo, Garnacha and Cabernet Sauvignon grapes from high- and low-quality vineyards over two years. *European Food Research and Technology 214,* 303–309.

Ayala, F. Echavarri J. F. and Negueruela A. I. (1997). A new simplified method for measuring the color of wines. I. Red and rose wines. *American Journal of Enology and Viticulture* 48, 357-363.

Boulton, R. (2001). The copigmentation of anthocyanins and its role in the color of red wine: A critical review. *American Journal of Enology and Viticulture 52*, 67–87.

Bravdo, B., Hepner, Y., Loinger, C., Cohen, S. and Tabacman, H. (1984). Effect of crop level on grown, yield and wine quality of a high-yielding Carignane vineyard. *American Journal of Enology and Viticulture 35*, 247–252.

Clingeleffer, P.R., Krstic, M.P. and Welsh, M.A. (2002). Effect of post-set, crop control on yield and wine quality of Syrah. Proceedings of the Eleventh Australian Wine Industry Technical Conference (The Australian Wine Industry Technical Conference Inc.: Urrbrae, South Australia) pp. 84–86.

Costa, G., Andreotti, C., Bucchi, F., Sabatini, E., Bazzi, C., Malaguti, S. and Rademacher, W. (2001). Prohexadione-Ca (Apogee): Growth regulation and reduced fire blight incidence in pear. *Horticultural Science 36*, 931–933.

Costa, G., Andreotti, C., Spinelli, F. and Rademacher, W. (2006). Prohexadione-Ca: more than a growth regulator for pome fruit trees. *Acta Horticulturae 727*, 107–116.

Del Rio, J. A., Baidez, A. G., Botia, J. M. and Ortuno, A. (2003). Enhancement of phenolic compounds in olive plants (Olea europaea L.) and their influence on resistance against Phytophthora sp. *Food Chemistry 83*, 75–78.

Diago, M.P., Vilanova, M., Blanco, J.A. and Tardaguila, J. (2010). Effects of mechanical thinning on fruit and wine composition and sensory attributes of Grenache and Tempranillo varieties (Vitis vinifera L.) *Australian Journal of Grape and Wine Research 16*, 314–326.

Disegna, E., Boido, E., Carrau, F., Fariña, L., Medina, K., Méndez, M., Rodríguez, P. and Dellacassa, E., (2003). Efectos de la aplicación del regulador de crecimiento 3,5-Dioxo-4-propionilciclohexancarboxilato de calcio (BAS 125) en la producción de uvas, composición del vino y aroma del cv. *"Tannat". Jornadas GESCO, 13., 2003,*. Libro de actas, 51-55, INIA, Montevideo, Uruguay.

Dubourdieu, D., Grassi,, C., Deruche, C. and Ribéreau-Gayon, P. (1984). Mise au point d'une mesure rapide de l'activité laccase, dans les moûts et dans les vins par méthode à la syringaldazine. Application à l'appréciation de l'état sanitaire des vendanges. *Connaissance Vigne Vin, 18, 237-252.*

Evans, J.R., Evans, R.R., Regusci, C.L. and Rademacher, W. (1999). Mode of action, metabolism and uptake of BASF 125 W prohexadione calcium. *HortScience 34*, 1200–1201.

Feucht, W. and Treutter, D. (1999). The role of flavan-3-ols and proanthocyanidins in plant defence. In Principles and Practices in Plant Ecology; Inderjit, D., Foy C. L., Eds.; CRC Press: Boca Raton, FL, pp 307–338.

Feucht, W., Treutter, D. and Schwalb, P. (1998). Principles of barrier formation of scab infected apple fruits. Z. *Pflanzenkr. Pflanzenschutz 105*, 394–403.

Fischer, T.C., Halbwirth, H., Roemmelt, S., Sabatini, E., Schlangen, K., Andreotti, C., Spinelli, F., Costa, G., Forkmann,G., Treutterc, D. and Stichb, K. (2006). Induction of polyphenol gene expression in apple (Malus x domestica) after the application of a dioxygenase inhibitor. *Physiologia Plantarum 128: 604–617.*

García-Escudero, E., Villar, M., Garcia-Oliveras, C., Ibáñez, S. and Romero, L. (2004). Influencia del aclareo de racimos en el rendimiento y calidad del vino en las variedades tintas de la D.O.Ca. Rioja. IV World Wine Forum, Logroño, Spain.

Gosch, C., Puhl, I., Halbwirth, H., Schlangen, K., Römmelt, S., Andreotti, C., Costa, G., Fischer, T. C., Treutter, D., Stich, K. and Forkmann, G. (2003). Effect of prohexadione-Ca on various fruit crops: flavonoid composition and substrate specificity of their dihydroflavonol 4-reductases. *European Journal of Horticultural Science 68*, 144–151.

Guidoni, S., Allara, P. and Schubert, A. (2002). Effect of cluster thinning on berry skin anthocyanin composition of *Vitis vinifera* cv. Nebbiolo. *American Journal of Enology and Viticulture 53*, 224–226.

Iland, P.G. (1987) Predicting red wine colour from grape analysis. *The Australian Grapegrower and Winemaker 285*, 29.

Iland, P.G., Bruer, N., Edwards, G., Weeks, S. and Wilkes, E. (2004). Chemical analysis of grapes and wine: tecniques and concepts. Campbelltown (Australia). Patrick Iland Wine Promotions PTY LTD,

Lo Giudice, D., Wolf, T.K. and Zoecklein B.W. (2004). Effects of prohexadione-calcium on grape yield components and fruit and wine composition. *American Journal of Enology and Viticulture 55*, 73–83.

Martinez de Toda, F. and Tardaguila, J. (2003). Meccanizzazione e fabbisogni di manodopera dei diversi sistemi di allevamento. In: Forme di allevamento della vite e modalitá di distribuzione dei fitofarmaci. Eds. P. Balsari and A. Scienza (Bayer Cropscience: Milan, Italy) pp. 143–158.

Mazza, G., Fukumoto, L., Delaquis, P., Girard, B. and Ewert, B. (1999). Anthocyanins, phenolics, and color of Cabernet Franc, Merlot, and Pinot Noir wines from British Columbia. *Journal of Agricultural and Food Chemistry 47*, 4009–4017.

McGrath, M. J., Koczan, J. M., Kennelly, M. M., and Sundin, G. W. (2009). Evidence that prohexadione-calcium induces structural resistance to fire blight infection. *Phytopathology 99*, 591–596.

Miller, S.S. and Tworkowski, T. (2003). Regulating vegetative growth in deciduous fruit trees. *Quarterly Reports on Plant Growth Regulation Society of America 31*, 8–46.

Owens, C.L. and Stover, E. (1999). Vegetative growth and flowering of young apple trees in response to prohexadione calcium. *Hortscience 34*, 1194–1196.

Parvez, M. M., Tomita-Yokotani, K., Fujii, Y., Konishi, T. and Iwashina, T. (2004). Effects of quercetin and its seven derivatives on the growth of Arabidopsis thaliana and Neurospora crassa. *Biochemical Systematics and Ecology 32*, 631–635.

Peña-Neira, A., Cáceres, A. and Pastenes, C. (2007). Low molecular weight phenolic and anthocyanin composition of grape skins from cv. Syrah (Vitis vinifera L.) in the Maipo valley (Chile): Effect of clusters thinning and vineyard yield. *Food Science and Technology International 13*, 153–158.

Petrie, P.R. and Clingeleffer, P.R. (2006). Crop thinning (hand versus mechanical), grape maturity and anthocyanin concentration: Outcomes from irrigated Cabernet Sauvignon (*Vitis vinifera* L.) in a warm climate. *Australian Journal of Grape and Wine Research 12*, 21–29.

Puhl, I., Stadler, F. and Treutter D. (2008). Alterations of flavonoid biosynthesis in young grapevine (*Vitis vinifera* L.) leaves, flowers, and berries induced by the dioxygenase inhibitor Prohexadione-Ca. *Journal Agriculture and Food Chemistry 56*, 2498–2504.

Rademacher, W. (2000). Growth retardants: effects on gibberellins biosynthesis and other metabolic pathways. *Annual Review of Plan Physiology and Plant Molecular Biology 51*, 501–531.

Rademacher, W and Kober, R. (2003). Efficient use of prohexadione-Ca in pomme fruits. *European Journal of Horticultual Science 68*, 101–107.

Römmelt, S., Fischer, T. C., Halbwirth, H., Peterek, S., Schlangen, K., Speakman, J.B., Treutter, D., Forkmann, G. and Stich, K. (2003b). Effect of dioxygenase inhibitors on the resistance-related flavonoid metabolism of apple and pears: Chemical, biochemical and molecular biological aspects. *European Journal of Horticultural Science* 68, 129–136.

Römmelt, S., Treutter, D., Speakman, J. B. and Rademacher, W. (1999). Effects of prohexadione-Ca on the flavonoid metabolism of apple with respect to plant resistance against fire blight. *Acta Horticulturae 489*, 359–364.

Römmelt, S., Zimmermann, N., Rademacher, W. and Treutter, D. (2003a). Formation of novel flavonoids in apple (Malus × domestica) treated with the 2-oxoglutarate-dependent dioxygenase inhibitor prohexadione-Ca. *Phytochemistry 64*, 709–716.

Rühmann, S. and Treutter, D. (2003). Effect of N-nutrition in apple on the response of its secondary metabolism to prohexadione-Ca treatment. *European Journal of Horticultural Science 68*, 152–159.

Sang-Mo Kanga, Jung-Tae Kimb, Muhammad Hamayuna, In-Cheon Hwangb, Abdul Latif Khana, Yoon-Ha Kima, Joon-Hee Leea and In-Jung Leea. (2010). Influence of prohexadione-calcium on growth and gibberellins content of Chinese cabbage grown in alpine region of South Korea. *Scientia Horticulturae 125*, 88–92.

Scalbert, A. (1991). Antimicrobial properties of tannins. *Phytochemistry 30*, 3875–3883.

Schwalb, P. and Feucht, W. (1998). Affinity of flavanols to cell walls: Changes after wounding, infection, lignification, and suberization. *Journal of Applied Botany and Food Quality 72*, 157–161.

Tardáguila, J., Petrie, P.R., Poni, S., Diago, M.P. and Martinez De Toda, F. (2008). Effects of mechanical thinning on yield and fruit composition of Tempranillo and Grenache grapes trained to a vertically shoot positioned canopy. *American Journal of Enology and Viticulture 59*, 412–417.

Vail, M.E. and Marois, J.J. (1991). Grape cluster architecture and the susceptibility of berries to *Botrytis cinerea*. *Phytopathology 81*, 188–191.

Vaquero-Fernández, L., Fernández-Zurbano, P., Sanz-Asensio, J., Lopez-Alonso, M. and Martínez-Soria, M.T. (2009). Treatment of grapevines with prohexadione calcium as a growth regulator. The influence on production, winemaking and sensory characteristics of wines. *Journal International des Sciences de la Vigne et du Vin 43*, 149–157.

Weber, B., Hoesch, L. and Rast, D. M. (1995). Protocatechualdehyde and other phenols as cell wall components of grapevine leaves. *Phytochemisty 40*, 433–437.

Yamamoto, M., Nakatsuka, S., Otani, H., Kohmoto, K. and Nishimura, S. (2000). (+)-Catechin acts as an infection-ihibiting factor in strawberry leaf. *Phytopathology 90*, 595–599.

Winkler, V.W. (1997). Reduced risk concept for prohexadione-calcium, a vegetative growth control plant growth regulator in apples. *Acta Hort, 451, 667-671.*

Yoder, K. S., Miller, S. S. and Byers, R. E. (1999). Suppression of fireblight in apple shoots by prohexadione-calcium following experimental and natural inoculation. *HortScience 34*, 1202–1204.

In: Grapes ISBN 978-1-61470-950-3
Editors: R. P. Murphy et al., pp. 155-170

Chapter 6

EXTRACTION AND BIOACTIVE PROPERTIES OF GRAPESEED POLYPHENOLS

Jorge Sineiro, Marivel Sánchez and María José Núñez
Department of Chemical Engineering. Technical School of Engineering.
University of Santiago de Compostela.
Santiago de Compostela. 15782. Spain

ABSTRACT

Grapeseeds have taken attention from the scientific community because of the high content of galloylated proanthocyanidins. The percentage of galloylated moieties respect to the total is noted as galloylation degree, which highly affects the antioxidant and anticarcinogenic activities. The antioxidant properties of proanthocyanidins from grape pomace and grapeseed as affected by processing and extraction conditions, is discussed in this chapter. Thermal treatments and the maintaining of a reducing medium by bubbling nitrogen have been reported to increase the antioxidant activity. The high positive correlation between polyphenol content and antioxidant activity was demonstrated, thus being the selection of solvent a key factor for maximizing the antioxidant activity. Ethanol:water mixtures have been found to be the best solvent. Galloylated proanthocyanidins from white varieties have been reported to be more active than those anthocyanins from the red ones. The main differences between red and white varieties are also discussed in this chapter.

1. Obtaining of Procyanidin-Enriched Fractions from Grapes

Grapeseeds have taken attention from the scientific community because of their high content of galloylated proanthocyanidins. Raw extracts have been traditionally obtained from grapeseeds using methanol [1, 2], methanol/water mixtures [3, 4] ethanol [5] and its mixtures with water [6] or acetone/water mixtures [4]. Alternatively, aqueous processes have been applied mainly to grape pomace [7, 8], for which extraction with water demonstrated several advantages and polyphenols extraction yield similar or higher than that of organic solvents. One of the main advantages of using ethanol instead of other organic solvents is its character of bio-renewable solvent, because it can be obtained by fermentation, thus with the potential of being an almost "zero-carbon-dioxide" emission product. The use of 1/1 (v/v) mixture with water, which has shown to be the best solvent for berry anthocyanins [9], has some advantages: the cost of solvent is lower, the process is safer and fractions can be obtained without lipids, even when grape pomace is used as raw extract; grape pomace contain easily-extractable waxes from skins, then requiring extraction with hexane to remove lipids [7].

Other alternative processes for polyphenols extraction from grapeseeds are those using supercritical CO_2 [10], but it has not been demonstrated they were economically competitive with solvent extraction, in some grade due to the need for adding polar solvents such as ethanol or methanol, to increase the polarity. Both pressurized liquid extraction and subcritical extraction are emerging extraction techniques, up to date studied mainly at laboratory scale [11]. Ultrasound-assisted extraction has been applied at laboratory scale [12], although is a promising technique because higher yields can be obtained for aqueous processes, which can be even increased by simultaneous enzymatic hydrolysis if yield was not high enough; moreover the necessary equipment for continuous extraction assays and scale-up is available.

Different yields and composition are expected when comparing grapeseeds from red and white varieties, because the first ones contain anthocyanins and also polymeric forms of anthocyanin-proanthocyanidin combinations [13], whereas the white ones contain proanthocyanidins and flavan-3-ols. The proanthocyanidin extraction yield depends mainly on the variety, being reported great differences when several varieties of red or white grapeseeds were compared [2]. In a study carried out on *Vitis vinifera* var. Shiraz (a red Variety), Kennedy et al. 2000 [14] reported that ripening highly influenced the profile of flavan-3-ol monomers and oligomers, diminishing the total content of monomers and the catechin/epicatechin ratio, together with the mean degree of polymerization of the oligomers. The effect of a previous processing is also important because wine-making procedures are different for red and white varieties: proanthocyanidins migrate partially to the red wine whereas white wines are not permitted to ferment in the presence of seeds and pomace [15]. Additionally, when grapes have been distilled for obtaining spirits, the extraction yield will likely be very different from that of non-distilled grapes. Although this process facilitates the extraction from seeds by increasing its porosity, an important fraction of polyphenols are extracted and did not remain in the wasted pomace [15] or the polyphenolic profile is modified by the thermal treatment [16].

2. Recovery of Proanthocyanidin-Enriched Fractions from Raw Extracts

Raw grapeseed extracts are nowadays in the market of nutritional supplements, often in form of capsules made of edible polymers, into which grapeseed oil was included as vehicle because it facilitated the extract handling and dosage.

These raw extracts are roughly characterized by the so-called "oligomer proanthocyanidins content" (OPC), which is an estimation of the content of oligomers, calculated in the basis of a HPLC chromatogram where the areas of monomers and polymers are removed and the remaining areas are expressed as percentage.

This parameter can be useful for comparative uses, but does not give information about the potential bioavailability and bioactivity of the compounds. Since many years, researchers have been studying the obtaining of fractions which content of oligomeric proanthocyanidins was best defined, expressing a "mean polymerization degree", often noted as mDP value for characterizing the size of oligomers, and a "galloylation degree", which is used to characterize the percentage of galloylated flavan-3-ols, such as catechin gallate or epicatechin gallate.

The techniques most often applied for obtaining short-range mDP value fractions involve gel filtration with selective media, usually applying a previous fractioning with solvents or pre-concentration in solid media. These previous stages are optional when raw extract is very rich in procyanidins, although it is recommended because a cleaner product and also a reduction of monomers can be achieved.

2.1. Preconcentration

The most often reported medium for pre-concentration of polyphenols are the styrene-divinylbenzene polymers from Röhm&Haas, named Amberlite® XAD, being Amberlite XAD16 that we prefer because its specific surface is higher than that of other polymers in the same series and it is not costly. When raw liquid extract is aqueous, its use is advantageous because it can retain selectively polyphenolic compounds, allowing sugars and proteins to be separated. Nevertheless, this is often an issue with raw extracts from grape pomace, but not when raw extracts are obtained from grapeseeds.

The necessity for removing ethanol by vacuum-evaporation for selectively retaining polyphenols in Amberlite XAD16 arises when using water/ethanol mixtures. This evaporation stage will be the only additional one, in contrast with several solvent extraction-evaporation stages that must be performed when proanthocyanidin-enriched fractions are being be obtained by using only liquid solvents.

2.2. Gel Filtration

The most popular processes for the recovery of proanthocyanidin oligomers involve gel chromatography as main stage. In this batch technique, samples are eluted through a column, a gradient of solvent mixtures being applied to sequentially elute several fractions depending on their size and/or polarity.

Several solid phases have being reported as suitable for the separation of proanthocyanidins according to their molecular weight, demonstrating different efficiencies. Probably, Toyopearl HW-40F was between the best ones, because it achieves a good separation according to molecular weight of oligomers. The most often reported solid phase is Sephadex LH-20, a dextrane crosslinked polymer designed to be lipophilic, thus showing acceptable chemical resistance for use with ethanol, methanol, etc. A drawback we have observed is that the separation of polyphenols in this material is a mixed phenomenon, where both polarity and molecular weight influence on the separation, thus rendering less chemically-defined fractions than Toyopearl for grape pomace extracts, for example. With Sephadex LH-20 is often to obtain oligomeric fractions containing some monomers, but it is efficient in removing polymers and highly polar compounds. For a similar proanthocyanidin-enriched extract, the peaks are less resolved than those obtained with Toyopearl HW-40F [17]. Both media are expensive, reason by which their use is nowadays limited to laboratory or pilot-plat scale obtaining of specific fractions for products testing.

An increasing interest has been focused on the silica-based solid media used in flash chromatography technique. The selectivity obtained through the use of spherical C18 beads, the same medium used in preparative and analytical application for polyphenols, is a great advantage, but nowadays these media are not economical beyond laboratory applications. Moreover, low-cost silicon coarse materials are also useful for polyphenols pre-concentration in low-scale processes.

It must be noted that these materials have being widely tested and used for purification of synthetic chemicals, being economical factors those limiting their use for natural products purification.

Gel filtration and flash chromatography automated equipment is available for medium-scale applications. Nevertheless, the development of processes for obtaining well chemically-defined proanthocyanidin fractions with proven bioactivity and at a cost low enough to be profitable for use as nutraceutical or food additive remains yet unsolved. The cost of chemically-defined fractions from grapeseeds instead of the extracts available in the market is prohibitive for the nutraceuticals or food supplements, but the important bioactivity of some specific fractions may change this situation.

3. Characterization of Grapeseed Proanthocyanidins

The composition of grapeseed extractable polyphenols has been most often determined using ethanol or ethanol/water mixtures as solvents. These ones show the highest potential for industrial processing because of its bio-renewable nature. Nevertheless, there are other possible choices, like pressurized acetone/water mixtures, which have been assayed at laboratory scale [18] or methanol/water mixtures [4].

With regards to grape proanthocyanidins, flavan-3-ol dimers are the most representative of these scompounds, nowadays being commercially available as procyanidin B1, B2, C1, etc. The techniques for separation and purification of dimers and trimers are time costly, involving several separation techniques, although advances in separation of oligomeric proanthocyanidins have been reported in recent years

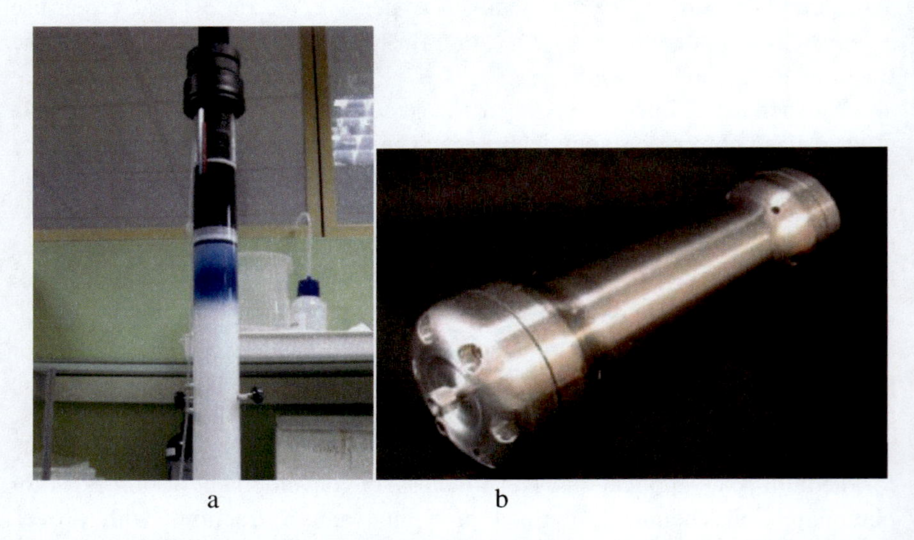

a b

Figure 1. a) View of 75x 100 mm column filled with Sephadex LH-20. B) Preparative cartridge for gel permeation filled with silica-based C18 stationary phase.

For example, in a recent study, Appledoorn et al. [3] applied selective solvent extraction, gel filtration, phase semi-preparative reversed-phase chromatography and normal-phase chromatography through a silica-based stationary phase to purify well-defined procyanidin fractions from grapeseed. In their work, dimers obtained from grape seeds were identified as epicatechin-(4-8)-catechin (B1), epicatechin-(4-8)-epicatechin (B2), catechin-(4-8)-catechin (B3) and catechin-(4-8)-epicatechin (B4).

Monomers (+)-catechin and (-)-epicatechin are the major monomeric flavonols in grapeseed, being present in most cultivars in larger amount than any of the procyanidin oligomers [3, 19] in most cultivars.

In most of the cultivars studied by these authors, B2 was the major proanthocyanidin and galloylated procyanidins were present in much lower concentrations than non-galloylated ones. Cultivar highly influences the flavan-3-ol composition of the seeds, being red varieties more reach in monomers, total proanthocyanidins and galloylated dimers [19, 20].

a) B-Type, 4-8 linkage b) B-Type, 4-6 linkage

Figure 2. Representation of B-type proanthocyanidin dimers.

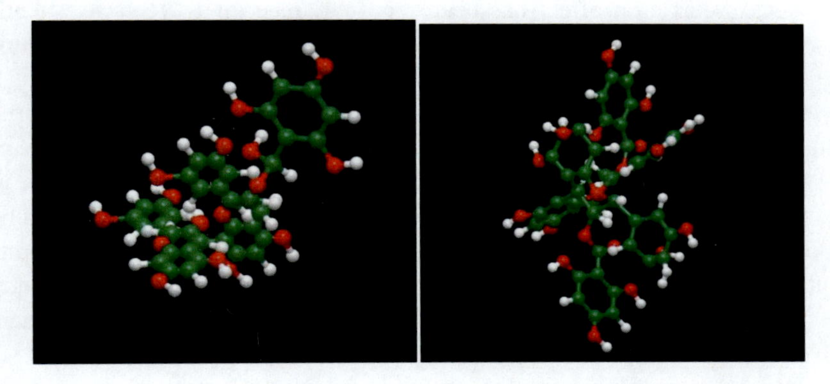

Figure 2. 3D ball-and-stick representation of B-type proanthocyanidin dimers : a) monogalloylated dimer (EC-ECG); b) digalloylated (ECG-ECG) dimers.

Several red varieties had a low content of galloylated proanthocyanidins, which indicates that not only the feature “red” or “white” of a grape variety should be taken into account for selecting raw materials when grape pomace is not a mix of varieties and the objective is to purify galloylated oligomers.

Gonzalez-Paramás et al. [21] analyzed the flavanols profile of extracts from grape pomace after several industrial processing that comprised distillation and further washing/drying stages. As they reported, dried grapeseeds from grape pomace still kept a high flavanol concentration and antioxidant activity, even though being subjected to thermal treatments. Proanthocyanidins have shown a thermal stability and oxidative stability in storage conditions that provides an advantage against other type of antioxidants such as retinoids or tocopherols. A main drawback for its use in foods is their well-known bitterness and astringency, due to their easy of binding to proteins.

Proanthocyanidins from grapes or grape extracts have been evaluated with regards to their antioxidant, cardioprotective and antitumoral activities since 90s, but most of the studies have been performed in the last ten years.

4. Bioactivity of Grapeseed Extracts And Proanthocyanidins

4.1. Vasorelaxing and Cardioprotective Activity

A vasorelaxation effect has been reported since many years in scientific papers with regards to a moderate consumption of red wine. The initial studies were motivated by the observation of the so-called “French paradox”: it is the lower incidence of atherosclerosis and cardiac events in France, in comparison with similar events in America, together with the fact that French diet was rich in saturated fats like those present in butter. Moderate wine consumption was identified as the main factor explaining this difference; several researchers reported in earlier papers that this healthy effect was due to the resveratrol content of wine, but nowadays the beneficial effects are also being studied with regards to the content of other polyphenols [22, 23] and not only that of resveratrol.

Major grape polyphenols are monomeric flavan-3-ols, anthocyanins and proanthocyanidins [24]. The simultaneous occurrence of anthocyanins and proanthocyanidins is also common in berries, which have been widely studied as source of antioxidant and cardioprotective substances, being blueberries and red grapes between those showing remarkable properties [25].

In vivo vasorelaxing activity of grapeseed proanthocyanidins has been recently studied by several authors through. Ortega et al.[26] studied the vasorelaxing activity of red wines on Wistar rats as affected by the grape variety; they found that the content of anthocyanins was the parameter most closely related with vasorelaxing activity, whereas monomeric flavanols (catechin, epicatechin) were not related. They did not make difference of galloylated and non-galloylated proanthocyanidins, which content was very low in wines, in comparison with grape pomace extracts. Fitzpatrick et al. [27]

related the high vasorelaxing effect of non-alcoholic grape skin and seeds extracts on pre-contracted vascular smooth muscles with galloylated procyanidins concentration, the effect increasing with the polymerization degree in the range studied. Galloylated oligomers, from dimers to tetramers, are those reported to be the most active compounds.

It has been also reported that oligomeric procyanidins relaxed human and animals vascular rings precontracted with noradrenaline and decreased vascular perfusion pressure [28].

4.2. Chemopreventive Activity on Atheroscleroses

Intake of red wine at moderate doses has been related with the reduction of atheroscleroses risk and cholesterol levels in vivo. Feeding extracts at a moderate dose, mimicking two glasses of red wine per meal, reduced plasma cholesterol (-11% on average), but did not affect plasma antioxidant capacity of hamsters [29]. The extracts prevented the development of aortic atherosclerosis by 68% (EGS), 63% (EGT), and 34% (GSE).

Caturla et al. [30] reported that in ex vivo experiments, galloylated catechins and especially ECG, dramatically affected the physical properties of both the phosphatidylcholine (PC) and phosphatidylethanolamine (PE) bilayers. These authors also report that galloylated catechins showed higher phospholipid/water partition coefficients than their, homologues.

4.3. Anticarcinogenic Activity

Proanthocyanidins have been studied *in vitro* on several cancer cells. The most widely studied have been colon cancer cells. It has been proven that oligomeric proanthocyanidins are not absorbed, thus being metabolized by colonic flora; either by the proper proanthocyanidins or through the action of their degradation products, the beneficial action can be exerted here, by inducing apoptosis or reducing the proliferative activity of cancer cells.

Differently than other sources of proanthocyanidins such as pine bark, grapeseed extracts contain galloylated proanthocyanidins, derived from catechin gallate and epicatechin gallate the majoritary monomers. The degree of galloylation has shown to play a major role in antioxidant and anti-carcinogenic activity. It has been reported that fractions with polymerization degrees between two and four and galloylation percentages about 25% were the most active antioxidants [31] and the most active anticarcinogenic on colon cancer cells. [32]. Galloylation was also reported recently to play an importante role in the anticarcinogenic activity of proanthocyanidins on prostate cancer [33].

Lizárraga et al. [32, 34] studied fractions with varying degrees of polymerization and galloylation on HT29 colon cancer cells, indicating the influence of electron transfer capacity on the antiproliferative and apoptotic activity. Roy et al [35] reported the induction of apoptosis on fibroblasts JB6 C141 through activation of p53 and Bax pathways. The main actions attributed to grapeseed proanthocyanidins, explaining their induction of apoptosis, are associated with increased expression of p53 protein and its phosphorylation to Ser^{15}, together with a downregulation of the antiapoptotic proteins Bcl-2 and Bcl-xl and overexpression of caspase-p, cleaved caspase-3, cytochrome C and pro-apoptotic protein Bax [35]. As main limitation, these authors indicate that, because apoptosis induction was developed through p53-dependent pathways, this chemopreventive action will not be observed in tumors expressing p53 mutants, which are very common in humans. More research must be done in this field.

Results of *in vivo* studies have been reported by Meeran et al. [36] and Katiyar [37], who concluded the existence of an association between grapeseed extract consumption and the decrease of UVB-induced skin cancer, in terms of incidence, multiplicity and malignant transformation of papillomas to carcinomas in mice. Activation of mitogen-activated protein kinases (MAPK), reduction of antioxidative stress and modification of cytokines has been observed after mice were fed grapeseed extracts [36, 37].

Lu et al. [38] studied the anti-angiogenic activity of grapeseed extracts through expression ofn vascular endothelium factor growth (VEGF), reporting that its mRNA and protein expression was inhibited in human glioma and breast cancer cells. They concluded that a block of Akt synthesis is the responsible for the reduction of hypoxia-inducible factor (HIF-1α) expression. Interestingly, it seems that grapeseed extracts suppressed the phosphorilation of several components involved in HIF-1a expression. Together with similar results on VEGF regulation *in vitro*, Wen et al. [39] also reported that GSE treatment of tumor bearing mice led to reduction of blood vessel density and phosphorylation of MAP kinase.

Another action promoted by grapeseed extract is the downregulation of Bcl-2 expression, while increasing the expression of Bax by grapeseed extracts in human epidermoid carcinoma A431 cells, followed by an increase of caspase-3 levels [36], which is a common observation for proanthocyanidins [33, 40].

The activation of NF-κB has been involved in inflammation, cell proliferation and oncogenic processes. It has been reported that grapeseed procyanidins exert their anti-cancer effects through the suppression of NF-κB [41].

Because these actions take place at cell level, bioavailability is a key factor in this action. Many of the results reported till date are not directly extrapolable because an extract was used in the in vitro assays, instead of small bioavailable fractions (containing dimers or, at most, trimers) or their metabolites.

Table 1. Chemopreventive actions reported by several authors in in vivo assays

Activity	System	Reference
Reducing VEGF growth	Mice breast cancer cells Human glioma cells/MDA-MB-231 human breast cancer cells	[39] [38]
MAPK regulation	Mice breast cancer cells	[39]
Increase of caspase-3/caspase-9 cleavage	Human prostate cancer LNCap cells Human keratinocytes JB6 C141	[33] [35]
PARP cleavage	Human prostate cancer LNCap cells	[33]
Caspase-3 activation	Human epidermoid A431 cancer cells	[36]
Increase of p53 tumor suppressor gene	Human keratinocytes JB6 C141	[35]
Reduce of expression of antiapoptotic Bcl and Bcl-xl and/or increase of Bax	Human prostate cancer LNCap cells Human epidermoid A431 cancer cells	[33] [36]
Cytokines TNF-α/ IL-10		[40]

VEGF: vascular endothelium growth factor; MAPK: Mitogen-activated protein kinases; PARP, poly (ADP-ribose) polymerase.

CONCLUSION

Grapeseeds provide a valuable source of galloylated proanthocyanidins, which are bioactive substances with potential as cardioprotective agents and chemoprotective against some type of carcinomas, although this last type of action was supported only by *in vitro* assays. Proanthocyanidin-enriched fractions can be obtained by solvent extraction, aqueous pressurized extraction, supercritical extraction and other processes that are nowadays under development. The cost of these fractions is relatively high because of the purifying stages necessary to obtain chemically-defined fractions instead of the extracts actually available in the market, which are claimed to contain oligomeric proanthocyanidins as major compounds.

Antiproliferative and apoptotic-inducing activities of grapeseed proanthocyanidins have been demonstrated in cell assays. It has been proven the reduction of oxidative stress in vivo, which is a risk factor for cardiovascular and carcinomas growth. Bioavailability is a key factor and, because positive effects have been proven *in vivo*, it has been suggested that the metabolites of proanthocyanidins released by the action of intestinal flora are responsible for the positive effects.

REFERENCES

[1] Escribano-Bailón, T.; Gutierrez-Fernandez, Y.; Rivas-Gonzalo, J. and Santos-Buelga, C. 1992. Characterization of Procyanidins of Vitis vinifera Variety Tinta del Pais Grape Seeds. *J. Agric. Food Chem. 1002, 40, 1794-1799.*

[2] Bordiga, M.; Travaglia, F.; Locatelli, M.; Coïsson, J.D.; and Arlorio, M. 2011. Characterisation of polymeric skin and seed proanthocyanidins during ripening in six Vitis vinifera L. cv. *Food Chem.*, 127 (2011) 180–187.

[3] Appeldoorn, MM.; Sanders, M; Vincken, JP; Cheynier, V; Le Guernevé, C; Hollman, PCH; Gruppen, H. Efficient isolation of major procyanidin A-type dimers frm peanut skins and B-type dimers from grape seeds. *Food Chem.*, 2009, 117, 713-720.

[4] Xu, C; Zhang, Y; Wanga, J; Lu, J.. Extraction, distribution and characterisation of phenolic compounds and oil in grapeseeds. *Food Chem.*, 2010, 122, 688–694.

[5] Torres, J. L.; Bobet, R. New flavanol-derivatives from grape (Vitis Vinifera) byproducts. Antioxidant aminoethylthio-flavan-3-ol conjugates from a polymeric waste fraction used as a source of flavanols. *J. Agric.* Food Chem.*, 2001, 49, 4627-4634.*

[6] Karvela, E.; Makris, D.P.; Kalogeropoulos, N., Karathanos, V.T. and Kefalas, P. 2010. Factorial design optimisation of grape (Vitis vinifera) seed polyphenol extraction. *Eur Food Res Technol.*, 2009, 229, 731–742.

[7] Sánchez, M; Franco, D; Sineiro, J; Magariños, B; Núñez, MJ. Antioxidant power, bacteriostatic activity, and characterization of white grape pomace extracts by HPLC–ESI–MS. *Eur. Food Res. Technol.,* 2009, 230 (2), 291-301.

[8] Maier, T; Göppert, A; Kammerer, DR; Schieber, A. and Carle, R. Optimization of a process for enzyme-assisted pigment extraction from grape (Vitis vinifera L.) pomace. *Eur. Food Res Technol.*, 2008, 227:267–275

[9] Karin J. Lopendia, Jorge Sineiro, Carolina Shene and Mónica O. Rubilar. 2009. Influence of the Heat Processing on the Antioxidant Activity of Bioactive Phenolic Compounds in Murta (Ugni molinae Turcz.) Juice. International Congress on Polyphenols. Salamanca, 2008.

[10] Murga, R; Ruiz, R; Beltrán, R. and Cabezas, JL. Extraction of Natural Complex Phenols and Tannins from Grape Seeds by Using Supercritical Mixtures of Carbon Dioxide and Alcohol. *J. Agric. Food Chem.*, 2000, 48, 3408-3412.

[11] Ju, ZY and Howard, LR. Effects of Solvent and Temperature on Pressurized Liquid Extraction of Anthocyanins and Total Phenolics from Dried Red Grape Skin. *J. Agric. Food Chem.*, 2003, 51, 5207-5213.

[12] 12 Ghafoor, K; Choi, YH; Jeon, YJ and; JO, IH. Optimization of Ultrasound-Assisted Extraction of Phenolic Compounds, Antioxidants, and Anthocyanins from Grape (Vitis vinifera) Seeds. J. Agric. Food Chem., 2009, 57, 4988–4994.

[13] Vidal, S; Meudec, E; Cheynier, V; Skourumiris; Hayasaka, Y. Mass Spectrometric Evidence for the Existence of Oligomeric Anthocyanins in Grape Skins. *J. Agric. Food Chem.*, 2004, 52, 7144-7151.

[14] Kennedy, JA; Troup, GJ; Pilbrow, JR; Hutton, DR; Hewitt, D; Hunter, CR; Ristic, R; Iland, PG; Jones, GP. Development of seed polyphenols in berries from Vitis vinifera L. cv. Shiraz. Austral. *J. Grape Wine Res.,* 2000, 6, 244-254.

[15] Pinelo, M; Rubilar, M; Jerez, M; Sineiro, J; Núñez, MJ. Effect of solvent, temperature and solvent-to-solid ratio on the total phenolic content and antiradicalary activity of extracts from different components of grape pomace. *J. Agric. Food Chem.*, 2005, 53, 2111-2117.

[16] Davidov-Pardo, G; Arozarena, I; Marín-Arroyo, MR. Stability of polyphenolic extracts from grape seeds after thermal treatments. Eur *Food Res Technol.*, 2011, 232, 211–220.

[17] Torres, JL; Varela, B; García, MT; Carilla, J; Matito, C; Centelles, J; Cascante, M; Sort, X. Valorization of Grape (Vitis vinifera) Byproducts. Antioxidant and Biological Properties of Polyphenolic Fractions Differing in Procyanidin Composition and Flavonol Content. *J. Agric. Food Chem.*, 2002, 50, 7548-7555.

[18] Kylli, P; Nohynek, L; Puupponen-Pimiä, R; Westerlund-Wikström, B; Leppänen, T; Jukka Welling, J; Eeva Moilanen, E; Heinonen, M. Lingonberry (Vaccinium vitis-idaea) and European Cranberry (Vaccinium microcarpon) Proanthocyanidins: Isolation, Identification, and Bioactivities. *J. Agric. Food Chem.*, 2011, 59, 3373–3384.

[19] Perez-Magariños, S. and González-San José, ML. Evolution of Flavanols, Anthocyanins, and Their Derivatives during the Aging of Red Wines Elaborated from Grapes Harvested at Different Stages of Ripening. J. Agric. *Food Chem.,* 2004, 52, 1181-1189.

[20] Fuleki, T. and da Silva, JMR. Catechin and Procyanidin Composition of Seeds from Grape Cultivars Grown in Ontario. *J. Agric. Food Chem.* 1997, 45: 1156-1160.

[21] González-Paramás, A; Esteban-Ruano, S; Santos-Buelga, C; de Pascual-Teresa, S; Rivas-Gonzalo, J.. Flavanol Content and Antioxidant Activity in Winery Byproducts. *J. Agric. Food Chem.*, 2004, 52, 234-238.

[22] Nishizuka, T; Fujita,Y; Sato, Y; Nakano, A; Kakino, A; Ohshima, S; Kanda, T; Yoshimoto, R; Sawamura, T.. Procyanidins are potent inhibitors of LOX-1: a new player in the French Paradox. *Proceedings of the Japan Academy*, Series B. 2011, 87, 104-113.

[23] Dell'Agli M; Buscialá, A; Bosisio, E. 2004. Vascular effects of wine polyphenols. *Cardiovasc Res.,* 2004, 63, 593-602.

[24] Monagas, M; Gomez-Cordovés, C; Bartolomé, B; Laureano, O; Da Silva, JM. Monomeric, Oligomeric, and Polymeric Flavan-3-ol Composition of Wines and Grapes from Vitis vinifera L. Cv. Graciano, Tempranillo, and Cabernet Sauvignon. *J. Agric. Food Chem.* 2003, 51, 6475-6481.

[25] Kristo, A.S; Kalea, AZ; Schuschke, DA; Klimis-Zacas, DJ. A wild blueberry-enriched diet (Vaccinium angustifolium) improves vascular tone in the adult spontaneously hypertensive rat. *J. Agric. Food Chem.*, 2010, 58, 11600-11605.

[26] Ortega, T.; De La Hera, E.; Carretero, E.; Gómez-Serranillos, P.; Naval, M.V.; Villar, A.M.; Prodanov, M.; Vacas, V.; Arroyo, T.; Hernández, T. and Estrella, I. 2008. InXuence of grape variety and their phenolic composition on vasorelaxing activity of young red wines. *Eur. Food Res Technol.* (2008) 227:1641–1650.

[27] Fitzpatrick, DF; Bing, B; Maggi, DA; Fleming, RC; O'Malley, RM. Vasodilating procyanidins derived from grape seeds. *Ann. NY Acad. Sci.,* 2002, 957, 78–89.

[28] Berti, F; Manfredi, B; Mantegazza, P; Rossoni, G. Procyanidins from Vitis vinifera seeds display cardioprotection in an experimental model of ischemia-reperfusion damage. *Drugs Exp. Clin. Res.*, 2003, 29, 207–216.

[29] Auger, C; Gérain, P; Laurent-Bichon, F; Portet, K; Bornet, AL.; Caporiccio, B; Cros, G; Teissédre, PL; Rouanett, JM. Phenolics from Commercialized Grape Extracts Prevent Early Atherosclerotic Lesions in Hamsters by Mechanisms Other than Antioxidant Effect. *J. Agric. Food Chem.* 2004, 52, 5297-5302.

[30] Caturla, N; Vera-Samper, E; Villalón, J; Mateo, R; Micol, M. The relationship between the antioxidant and the antibacterial properties of galloylated catechins and the structure of the phospholipid model membranes. *Free Rad. Biol. Medicine.*, 2003, 34, 648–662.

[31] Pazos, M; Gallardo, JM; Torres, JL; Medina, I. Activity of grape polyphenols as inhibitors of the oxidation. of fish lipids and frozen fish muscle. *Food Chem.*, 2005, 92, 547–557.

[32] Lizárraga, D; Touriño, S; Reyes-Zurita, FJ; de Kok, TM; van Delft, JH; Maas, LM; Briedé, JJ; Centelles, JJ; Torres, JL; Cascante, M. 2008. Witch hazel (Hammamelis virginiana) fractions and the importance of gallate moieties--electron transfer capacities in their antitumoral properties. *J. Agric. Food Chem.*, 56, 11675-11682.

[33] Chou, S-C; Kaur, M; Thompson, JA.; Agarwal, R; Agarwal, C. Influence of Gallate Esterification on the Activity of Procyanidin B2 in Androgen-Dependent Human Prostate Carcinoma LNCaP Cells. *Pharm. Res.,* 2010, 27, 619-627.

[34] Lizárraga, C; Lozano, C; Briedé, JJ; van Delft, JH; Touriño, S; Torres, JL; Cascante, M. The importance of polymerization and galloylation for

the antiproliferative properties of procyanidin-rich natural extracts. *FEBS Journal*, 2007, 274, 4802-4811.

[35] Roy, AM; Baliga, MS; Elmets, CA; Katiyar, S. Grape Seed Proanthocyanidins Induce Apoptosis through p53, Bax, and Caspase 3 Pathways. *Neoplasia,* 2005, 7 , 24-36.

[36] Meeran, SM and Katiyar, SK. Grape seed proanthocyanidins promote apoptosis in human epidermoid carcinoma A431 cells through alterations in Cdki-Cdk-cyclin cascade, and caspase-3 activation via loss of mitochondrial membrane potential. *Exp. Dermatol.*, 2007;16:405–415.

[37] Katiyar, S.K. Grape seed proanthocyanidines and skin cancer prevention: Inhibition of oxidative stress and protection of immune system. *Mol. Nutr. Food Res*. 2008, 52(Suppl 1), S71–S76.

[38] Lu, J; Zhang, K; Chen, S.; Wen, W. Grape seed extract inhibits VEGF expression via reducing HIF-1a protein expression. *Carcinogenesis,* 2009, 3, 636–644.

[39] Wen, W; Lu, Y; Zhang, K; Chen, S. Grape seed extract (GSE) inhibits angiogenesis via suppressing VEGFR signaling pathway. *Cancer Prev. Res.*, 2008, 1, 554–561.

[40] Nadakumar, V; Singh, T; Katiyar, SK. Multi-targeted prevention and therapy of cancer by proanthocyanidins. *Cancer Lett.,* 2008, 269, 378-387.

[41] Mantena,SK and Katiyar, SK. 2006. Grape seed proanthocyanidins inhibit UV-radiation-induced oxidative stress and activation of MAPK and NF-kappaB signaling in human epidermal keratinocytes. *Free Radic Biol* Med., 40: 1603–1614.

In: Grapes ISBN 978-1-61470-950-3
Editors: R. P. Murphy et al., pp. 171-181 ©2012 Nova Science Publishers, Inc.

Chapter 7

ALLERGIES TO GRAPES

N. Sebastià[a], J. M. Soriano[a] and I. Gavidia[b]
[a]Department of Preventive Medicine
[b]Department of Plant Biology.
Faculty of Pharmacy. University of Valencia.
Av. Vicent A. Estelles s/n. 46100-Burjassot. Spain

ABSTRACT

Although allergies to grape are uncommon, allergic reactions to different grapes or grape-derived products have been described in the last two decades. Several allergens have been identified including endochitinase 4A, with a molecular mass around 30 kDa, a lipid-transfer protein (LTP) homologous to and cross-reactive with peach LTP, and a 24 kDa protein homologous to the cherry thaumatin-like allergen. The presence of these allergens in grape berries could explain the cross-reactive allergy to other foods in patients with an allergy to grapes. For a variety of reasons, allergy to grapes is peculiar as some patients only present allergy to one grape variety, suggesting that this allergy may be specific to a particular variety. On the other hand, other allergic patients may not tolerate any grape species, wine or raisins. Another more striking point is allergy to grapes only becomes apparent in some cases when fruit ingestion is combined with other concurrent factors such as exercise or alcohol.

This chapter reviews several case reports on grape allergy and places special emphasis on the diversity of the factors involved in this hypersensitive response to grape.

Introduction

The European Academy of Allergy and Clinical Immunology [1] proposed a mechanistic classification of food allergies:

i) *Adverse food reactions* are defined as any aberrant reaction after the ingestion of a food or food additive which may be the result of toxic or nontoxic food reactions.
ii) *Toxic reaction*s can occur in anyone, provided a sufficient dose is ingested (eg, histamine in scombroid fish poisoning).
iii) *Nontoxic reactions* depend on individual susceptibilities and may be the result of immune mechanisms (allergy or hypersensitivity) or nonimmune mechanisms (intolerance).
iv) *IgE-mediated food allergies* have been most clearly delineated, but non-IgE-mediated immune reactions, especially of the gastrointestinal tract, are increasingly recognized.
v) *Food intolerances* probably account for the majority of adverse food reactions and may be caused by the food's pharmacologic properties (eg; headaches from tyramine in aged cheeses and jitteriness from caffeine in coffee or soft drinks), or by the host's unique susceptibilities, such as metabolic disorders (eg; lactase deficiency) or idiosyncratic responses.

Nowadays, the prevalence of food allergy is around 3% in the adult population and 6-8% in young children [2]. A large number of plant food allergens has so far been identified [3], and most present a cross-reactivity between plant foods and pollens, or different plant sources (eg; fruits and latex) [4]. These allergens can be assigned to only 31 different protein families, and the majority is included in a few types of seed storage or plant-defense proteins [5]. These allergens can be grouped into three types of the panallergen family constituted by profilins, homologous to Bet v 1 (a major birch pollen allergen), and plant nonspecific lipid transfer proteins (ns-LTPs) [6].

i) *Profilins* are the proteins involved in the regulation of the actin cytoskeleton and are ubiquitous allergens. They have been identified in many plant foods as being responsible for cross-reactivity with pollens [7]. In fact, human IgE-recognizing pollen profilins are highly cross-reactive to profilins in cherry, pear, apple, celery, soybean, peanut, hazelnut, tomato and bell pepper [8].

ii) *Group of defense proteins homologous to Bet v 1* which behave as major allergens in patients from northern and central Europe with an allergy to vegetables associated with birch pollen allergy [6].

iii) Nonspecific lipid transfer proteins (ns-LTPs) are divided into families according to their molecular masses; 9-kDa ns-LTP1 and 7-kDa ns-LTP2. ns-LTP1 allergens have a basic pI, a conserved cysteine residue, and form a multigenic family whose genes are spatially and temporally regulated [9, 10]. They have been identified as major allergens in fruits from the Rosaceae family. However given their ubiquitous character, they are widely distributed in other plant organs. These proteins are resistant to heat denaturation and proteolysis by digestive enzymes [11]; hence, their importance as phytoallergens. Currently, an increasing number of LTP-homologous proteins, with a similarity of more than 80%, have been identified in several plant foods, particularly stone fruits, but also blueberry and raspberry [8].

Allergy to Grapes

According to the International Organization of Vine and wine (OIV) [12], in 2007 there were 7 792 000 hectares of grape growing across the planet and the world grape production reached approximately 200 M quintals. The vast majority is farmed for commercial purposes and originates from a species called *Vitis vinifera*, specifically from the many thousands of varieties within this species, i.e. Chardonnay, Cabernet-Sauvignon, Pinot noir, etc. Humans consumed this immense amount of grape as fresh grape, grape juice, wine or raisins.

Despite the vast areas of *Vitis vinifera* cultivation and the consumption of this fruit in the Mediterranean area, allergy to grape is rather uncommon. Indeed, the first case of grape allergy was reported by Senff et al. in 1990 [see 13]. The literature includes some reports of allergic reactions to all the different grape varieties and grape by-products, and even cases of anaphylactic reactions caused by eating grapes. An allergic reaction to different grape varieties could mean that grape allergy may be specific to a certain grape variety with tolerance to the others. However, patients may sometimes not tolerate any grape species, wine or raisins at all [14-16].

Some allergens of grape have been identified; the mayor allergen is an endochitinase 4A, with a molecular mass of around 30 kDa, followed by a lipid-transfer protein (LTP) homologous to peach LTP, and by a 24 kDa protein homologous to the cherry thaumatin-like allergen, which is a minor allergen [17, 18].

Although allergy to grapes is unusual, its research is good owing to several facts; for instance, several case reports have related it with food cross-reacting allergies: allergy to white or red wine, grass, olive and *Parietaria*, *Aspergillus*, Rosaceae family (apricot, blackberry, peach, cherry, apple, almond), banana, kiwi and latex [13]. Several manifestations could appear (asthma, anaphylactic shock) instead of the most common oral allergy syndrome (OAS) [13]. Moreover, and in many cases, diagnosing the allergy is no easy task; that is, immunologic tests, prick tests or oral challenges do not necessarily share the same results. Meanwhile, oral food challenges continue to be the gold standard in the diagnostic workup. Hence, each case report of grape allergy must be analyzed individually with a thorough medical history-taking and physical examination, and with laboratory tests used as important adjunct tools to confirm the diagnosis and to monitor its course [19].

We aim to describe several case reports which enclose the diversity of factors involved in this hypersensitive response to grape.

1. Grapes Allergy Including Oral Allergy Syndrome (OAS) and Other Allergic Symptoms

OAS is characterized by tingling, pruritus, erythema or angio-edema of the lips, tongue and soft palate after the oral cavity comes into contact with a variety of plant-derived foods, but not after ingestion. Sometimes, however, it may also include the gastrointestinal, respiratory or even cardiovascular systems.

Quite lot patients with grape allergy have suffered this syndrome after eating grapes or drinking red or white wine. In a study of Guinnepain and Rossemont [20], two patients first developed OAS to grapes and one of them experienced anaphylactic shock at the end of a meal. Furthermore, prick-by-prick tests were positive for red grapes, specific IgE were found and the subjects were also sensitized to pollens (mugwort and birch, respectively) and to the Rosaceae family. From this report, the authors suggested taking into account the risk of anaphylaxis after ingesting grapes.

A similar case was reported by Giannocaro et al. [15] in a woman who experience similar symptoms (flushing of the face and neck, with an itchy skin rash, edema of the oral and peroral mucosa, and dyspnea) which emerged 10 min after eating fresh grapes. The patient presented hypersensitivity to fresh grapes in a skin-prick test, and also to some pollens, plums and cherries.

Pastorello et al. [17] reported several cases of patients with allergy to grapes or wine mixed with other allergies and symptoms. Patients with allergy to wine also gave positive results for fresh grape in skin-prick tests, but were allergic only to specific young wines since they tolerated normal red wine. On the other hand, patients with a history of OAS after eating grapes changed their symptoms to laryngeal edema. They reported that the main allergen was a 30 kDa protein to which most patients reacted. It was identified as an endochitinase 4A. Furthermore, they found a protein belonging to the family of Lipid Transfer Proteins (LTP) that was homologous to and cross-reactive with peach LTP. A 24 kDa protein homologous to the cherry thaumatin-like allergen was a minor allergen. Which might explain the cross-reactivity of some patients with grape allergy to other foods, and others are monosensitized to grapes.

2. Food-Dependent Exercise-Induced Anaphylaxis

Food-dependent exercise-induced anaphylaxis, a particular form of anaphylaxis caused by physical exercise and ingestion of food, was described at the beginning of the 1980's. It comprises a heterogeneous group, with or without food allergy and probably some intermediary forms [21]. Grapes are included among the foods most often responsible for this anaphylactic reaction, along with seafood, celery, tomato, apple, hazel nut, orange, peach or cabbage.

In a study done with eleven Japanese patients [22], seven experienced anaphylactic symptoms only after eating certain foods, such as shellfish, wheat or grape before exercising. Strangely, those patients who developed anaphylactic symptoms before the age of 20 were atopic themselves or had atopic first-degree relatives and IgE antibodies against the causative food allergens were found in most of them. This situation did not occur in patients who developed anaphylaxis beyond the age of 30.

In Italy, a similar case was reported by Senna et al. [23]. This time, the patient had a medical history consisting in rhinoconjunctivitis due to *Parietaria* and mugwort, with facial flushing, edema of the lips and dyspnea after drinking wine, although she could eat fresh grapes. Nevertheless, on one occasion she ate grape and went jogging; afterward she suffered urticaria, facial/pharyngeal edema, abdominal pain and dyspnea, and she rapidly worsened. The prick-to-prick tests with juice from white and red grapes gave positive results. Even when no grape-specific IgE could be found in the patient's serum, IgE reactivity of serum against grape extract was confirmed. The authors confirmed that the case therein described was a true food-dependent exercise-induced anaphylaxis since the patient had no effects when eating large amounts of fresh grapes.

In all cases, some authors conclude that prevention is based on easy measures: I) observe the rule of 3 hours between meals and exercise, II) avoid physical efforts at high temperatures, III) pay attention to hidden food in food and energy products for athletes, IV) avoid drugs before physical effort, and IV) drink abundantly which exercising [21].

3. Alcohol-Induced Anaphylaxis to Grapes

In the literature we found several case reports covering allergy to grapes, and we encountered one of its alcohol by-products, wines, at the same time [24-26]. This is likely to be due to the same allergens being present in both products [17].

However, we found case reports in which allergy to grapes is induced by an alcoholic product. Alcoceba-Borrás et al. [18] reported a case report of a young woman who remained asymptomatic when eating grapes alone or when drinking alcohol, but presented anaphylactic reactions when eating grapes and alcohol together. Moreover, alcoholic beverages are a grape by-product. The patient presented generalized urticaria, facial angioedema and nasal obstruction twenty minutes after eating twelve grape grains and a glass of champagne. Nonethelesss she presented good tolerance to champagne, red wine, white wine, vinegar, red grape, white grape and muscatel grape individually. The prick-by-prick test was positive for red grape and white grape, but was negative for champagne. Since the patient did not present allergy when eating grapes alone, the authors decided to do an oral challenge with grapes and champagne. No adverse reaction was produced when a small quantity of both was administered. Hence, they reproduced the situation of combining twelve grape grains and 50 cc of champagne. Twenty minutes later the patient presented the symptoms she had complained about before. Finally, she avoided mixing these two products. Although the patient had given positive skin test results for grapes, no grape-specific IgE was found; obviously, champagne was a cofactor in the anaphylactic reaction.

4. Grape Allergy in Children

It seems that grape allergy is most common in adults than in children; in fact in the early 21st century [13], only three cases in children and about twenty in adults have been reported. Reported symptoms in children included angio-edema, urticaria and OAS in a syndrome of multiple allergies.

A case report of syndrome of multiple allergies was described in a child who had suffered episodes of urticaria during breastfeeding, and had also had these episodes at 2, 6, 10 and 20 months of age caused by several foods. Then at the age of four, an oral allergy syndrome after consuming red grapes, kiwi, coconut and almond was manifested. Overall, this child was diagnosed with multiple allergy syndrome to mites, various pollens of trees and grass, latex, kiwi, grapes, egg and fish.

On the other hand, another case report [13], in which the child had been previously diagnosed with atopic dermatitis, reported sub-clinical sensitization to grass and olive, and urticaria-angio-edema caused by nuts, apple, kiwi, peach and egg. Although the patient had tolerated grapes in the past, and allergy syndrome with lower-lip angio-edema developed only minutes after eating fresh grape species (mainly the Moscatel variety of white grapes). In fact, the IgE patient's serum recognized a differently sized protein (of 94,000 MW) in the Moscatel extract. In another study with 14 children who had suffered allergic reactions involving fresh grapes or grape juice, the same author affirmed that the prick-to-prick technique with pulp and/or peel from fresh grape could provide better results than the skin-prick test with commercial extracts. This study also stated that the skin-prick test gave a better predictive value than specific IgE levels because specific IgE grape levels found in many cases are lower than those mentioned in food allergens [27].

Indeed, a study of Cardinale et al. [28] showed that proteins are responsible for allergic reactions to a great extent. A boy with anaphylaxis to white fresh grapes who gave positive results in the skin-prick test for grape skin, grape juice and grape-specific serum IgE, did not report an allergic reaction when using an oil bath which contained grape seeds. Actually, the complete refining of oils almost totally removes proteins from oil.

5. Grape Allergy in Old-Aged People

Hypersensitivity to a commonly consumed fruit such as grapes can develop late in life causing near-fatal anaphylaxis [29]. The case report of a 66-year-old man who had tolerated red wine and grape with no previous reaction until 2003 is stressed. However, one year after, he developed an episode of anaphylaxis with a swelling of the tongue and eyes, and respiratory distress two hours after eating grapes.

Later in 2007, the patient had an episode of anaphylaxis with a swelling of the tongue and respiratory distress two hours after consuming red wine. The only medical history consisted in a 10-year history of seasonal rhinitis during the birch and grass pollen season. Finally, the skin-prick test gave positive results for red wine, white wine and grapes, and the patient's serum specific IgE indicated a reaction against birch, mug word, egg, milk, codfish, wheat flour, soy, peanut, apple, carrot, latex, grape, celery, cherry, and peach.

The authors concluded that a history of allergic reactions or positive sensitization towards peach and cherry should be taken seriously and should form part of the diagnostic workup for patients with an allergy to grape and wine [26].

6. Allergy to Vine Pollen and Grape

The first case of grape allergy in patients who are allergic to vine pollen was reported by Mur et al. [30] in 2006. A young woman who had suffered seasonal rhinoconjunctivitis with sensitization to pollens from vine, grass, olive, and Chenopodiaceae plants for years experienced two episodes of generalized itching, a maculopapular rash and facial angio-edema only a few minutes after eating grapes.

The skin-prick tests revealed positive reactions with extracts from the *Vitis vinifera* pollen and commercial grapes, while the prick-by-prick tests revealed a reaction with grapes.

Positives reactions to other fruits were also revealed, such as peach, banana, kiwi, and cherry. The clinical history and positive results to cutaneous and conjunctival tests to vine pollen extracts suggest that vine pollen was the primary sensitizer in this patient, who did not suffer from fruit allergy because the test showed no low-molecular-weight-binding proteins.

REFERENCES

[1] Bruijnzeel-Koomen, C.; Ortolani, C.; Aas, K.; Bindslev-Jensen, C.; Björkstén, B.; Moneret-Vautrin, D.; Wüthrich, B. Adverse reactions to food. *Allergy,* 1995, 50, 623-35.

[2] Burks, W.; Ballmer-Weber, B.K. Food allergy. *Molecular Nutrition and Food Research*, 2006, 50, 595-603.

[3] Breiteneder, H.; Radauer, C. A classification of plant food allergens. *Journal of Allergy and Clinical Immunology*, 2004, 113, 821-830.

[4] Van Ree, R. Clinical importance of cross-reactivity in food allergy. Current Opinion in Allergy and Clinical Immunology, 2004, 4, 235-240.

[5] Chapman, M.D.; Pomes, A.; Breiteneder, H.; Ferreira, F. Nomenclature and structural biology of allergens. *Journal of Allergy and Clinical Immunology,* 2007, 119, 414-420.

[6] Fernández-Rivas, M. Cross-reactivity between fruit and vegetables. *Allergologia et immunopathologia*, 2003, 3, 141-146.

[7] Hofmann, A.; Burks, A.W. Pollen food syndrome: update on the allergens. Review. *Current Allergy and Asthma Reports*, 2008, 8, 413-417.

[8] Marzban, G.; Mansfeld, A.; Hemmerb, W.; Stoyanova, E.; Katinger, H.; da Câmara Machado, M.L. Fruit cross-reactive allergens: A theme of uprising interest for consumers' health. *BioFactors,* 2005, 23, 235-241.

[9] Kader, J.C. Lipid-transfer proteins in plants. *Annual Review of Plant Physiology and Plant Molecular Biology*, 1996, 47, 627-654.

[10] Dubreil, L.; Gaborit, T.; Bouchet, B.; Gallant, D.J.; Broekaert, W.F.; Quillien, L.; Marion, D. Spatial and temporal distribution of the major isoforms of puroindolines (puroindoline-a and puroindoline-b) and non-specific lipid transfer protein (ns-LTP1) of Triticum aestivum seeds. Relationships with their in vitro antifungal properties. *Plant Science,* 1998, 138, 121–135.

[11] Rougé, P.; Borges, J.P.; Culerrier, R.; Brulé, C.; Didier, A.; Barre, A. Lipid transfer proteins as relevant fruit allergens. *Revue Française d'Allergologie,* 2009, 49, 58–61.

[12] International Organisation of Vine and Wine. Available from: http://www.oiv.int/uk/accueil/.

[13] Pétrus, M.; Malandain, H. Food allergy to grape. A new observation in a four years old child. Revue française d'allergologie et d'immunologie *Clinique,* 2002, 42, 806–809.

[14] Rodríguez, A.; Trujillo, M.J.; Matheu, V.; Baeza, M.L.; Zapatero, L.; Martínez, M. Allergy to grape: a case report. *Pediatric Allergy Immunology*, 2001, 12, 289-290.

[15] Giannocaro, F.; Munno, G.; Rivas, G.; Pugliese, S.; Paradiso, M.T.; Ferrannini, A. Oral allergy syndrome to grapes. *Allergy,* 1998, 53, 451-452.

[16] Bircher, A.J.; Bigliardi, P.; Yilmaz, B. Anaphylaxis resulting from selective sensitization to Americana grapes. *Journal of Allergy and Clinical Immunology*, 1999, 104, 1111-1113.

[17] Pastorello, E.A.; Farioli, L.; Pravettoni, V.; Ortolani, C.; Fortunato, D.; Giuffrida, M.G.; Garoffo, L.P.; Calamari, A.M.; Brenna, O.; Conti, A. Identification of grape and wine allergens as an endochitinase 4, a lipid-transfer protein, and a thaumatin. *Journal of Allergy and Clinical Immunology,* 2003, 111, 350-59.

[18] Alcoceba-Borràs, E.; Faraudo, E.B.; P.G. Jané.; Zavala, B.B. Alcohol-induced anaphylaxis to grape. *Allergologie et Immunopathologie* 2007, 35, 159-61.

[19] Gerez, I.F.A.; Shek, L.P.C.; Chung, H.H.; Lee, B.W. Diagnostic tests for food allergy. *Singapore Medical Journal*, 2010, 51, 4-9.

[20] Guinnepain, M.T.; Rassemont, R.; Claude, M.F.; Laurent, J. Oral allergy syndrome (OAS) to grapes. *Allergy,* 1998, 53, 1225.

[21] Dutau, G.; Rancé, F. Exercice and food-induced anaphylaxis. Revue française d'allergologie et d'immunologie *Clinique,* 2007, 47, S47-S54.

[22] Dohi, M.; Suko, M.; Sugiyama, H.; Yamashita, N.; Tadokoro, K.; Juji, F.; Okudaira, H.; Sano, Y.; Ito, K.; Miyamoto, T. Food-dependent, exercise-induced anaphylaxis: A study on 11 Japanese cases. *Allergy and Clinical Immunology*, 1991, 87, 34-40.

[23] Senna, G.; Mistrello, G.; Roncarolo, D.; Crivellaro, M.; Bonadonna, P.; Schiappoli, M.; Passalacqua, G. Exercise-induced anaphylaxis to grape. *Allergy,* 2001, 56, 1235–1236.

[24] Schäd, S.G.; Trcka, J.; Vieths, S.; Scheurer, S.; Conti, A.; Bröcker, E.B.; Trautmann, A. Wine anaphylaxis in a german patient: IgE-mediated allergy against a lipid transfer protein of grapes. *International Archives of Allergy and Immunology*, 2005, 136, 159-164.

[25] Kalogeromitros, D.C.; Makris, M.P.; Gregoriou, S.G.; Mousatou, V.G.; Lyris, G.; Tarassi, K.E.; Papasteriades, C.A. Grape anaphylaxis: a study of 11 adult onset cases. *Allergy and Asthma Proceedings, 2005, 26, 53-58.*

[26] Sbornik, M.; Rakoski, J.; Mempel, M.; Ollert, M.; Ring, J. IgE-mediated type-I-allergy against red wine and grapes. *Allergy,* 2007, 62, 1339 – 1348.

[27] Rodríguez, A.; Matheu, V., Trujillo, M.J., Martínez, M.I.; Baeza, M.L.; Barranco, R.; de Frutos, C.; Zapatero, L. Grape allergy in paediatric population. *Allergy,* 2004, 59-364.

[28] Cardinale, F.; Berardi, M.; Chinellato, I.; Damiani, E.; Nettis, E. A child with anaphylaxis to grapes without reaction to grape seed oil. *Allergy,* 2010, 65, 791–804.

[29] Vaswani, S.K.; Chang B.W.; Carey R.N.; Hamilton, R.G. Adult onset grape hypersensitivity causing life threatening ananphylaxis. *Annals of allergy asthma and immunology,* 1999, 83, 25-26.

[30] Mur, P; Brito, F.F.; Bartolomé, B.; Galindo, P.A.; Gómez, E.; Borja, J.; Alons, A. Simultaneous allergy to vine pollen and grape. *Journal of Investigational Allergology and Clinical Immunology,* 2006, 16, 271-273.

In: Grapes ISBN 978-1-61470-950-3
Editors: R. P. Murphy et al., pp. 183-192

Chapter 8

GRAPES AS AN ALTERNATIVE CROP FOR WATER SAVING

Khaldoon A. Mourad and Ronny Berndtsson
Department of of Water Resources Engineering,
Lund University, Lund, Sweden

ABSTRACT

Fruit trees have a vast range of water needs. When it comes to Crop Water Requirement (CWR), grapes may be considered as a low water consumption crop. Thus, grapes can be a good alternative in arid and semiarid areas as compared to dates, citrus, and bananas that have higher CWR. Much water can be saved if agricultural management focuses on high-yield crops with low CWR. Therefore, changing existing water-wasting practices from high to lower CWR crops can save water and improve the virtual water (embedded water) balance for water-scarce countries. This chapter estimates and discusses the potentially saved water amount from changing crop pattern into grapes in Syria and Jordan by computing the embedded water in different typical crops. The results can be used to better manage scarce water resources and lead forward to sustainable water management. This is especially important in the Middle East that faces rapidly depleted renewable water resources.

Keywords: crop pattern, integrated water resources management, virtual water, Jordan, Syria.

1. INTRODUCTION

Water poor countries in Middle East and North Africa (MENA) region have to reallocate agricultural water for more sustainable use of scarce water resources. For a sustainable agriculture, governments should base their agricultural plans on water productivity, virtual water, and crop water requirement (CWR), acknowledging that these dimensions can play a vital role in a warmer climate and with a larger population.

The embedded water in a product is called virtual water. Virtual water trade is a way that water can cross borders without being affected by political intentions or conflicts. This can be considered as an international water trade. Therefore, water-poor countries can balance water scarcity at their national level by importing high water content products and exporting low water content products. Estimating the virtual water in a product means including all embedded water that were used in its producing process. This means, for an agricultural product, including water from rainfall, fertilizers, pesticides, and other types of indirect water usage. However, some of the indirect water usage may be difficult to quantify. Virtual water content can be estimated from CWR as suggested by Mourad et al. (2010).

CWR is the water needed for the crop during its life cycle. CWR depends mainly on climate conditions, namely temperature, evaporation, humidity, and rainfall. For the same crop, CWR is higher in arid and semiarid areas as compared to humid areas. Estimating CWR must be based on experiments, field work, and research. According to Shatanawi et al. (1998) grape water requirement in Jordan is about 4500 m^3/ha (Figure 1), which is similar as for Syria (MAAR, 2010). As shown in Figure 1, planting figures, olives, and grapes can save water as compared to apples, bananas, and dates (Santesteban et al., 2011).

1.4. Grape Cultivation

Grape/vine needs specific climatic conditions;

- Temperature: Grapes need cold winters with low temperature (2–10°C) during at least two months. Temperature below zero can cause serious damage to grape cultivation especially in a long term. On the other hand, grapes need a long warm growing season (25–30°C) to complete maturity. However, high temperature (more than 38°C) can damage its vegetation growth.

- Humidity: High air humidity at flowering season can cause flower drop-off, which is noted in areas with fog and clouds. Furthermore, high humidity during the summer growing season can spread fungal diseases.
- Rainfall and irrigation: For rainfed agriculture, grapes need about 500 mm annually to grow without irrigation. However, if rainfall is not enough, irrigation is needed especially during dry months (June, July, and August). Irrigation, on the other hand, increases yield, vine water status, vegetative growth, and vine evapotranspiration (e.g., Intrigliolo and Castel, 2007). Irrigation also increases sugar content in the grapes and rises pH in wine later on, which affect the wine quality (Intrigliolo and Castel, 2009). Deficit irrigation, reducing canopy vigor, increasing fruit exposure to light and reducing fruit growth to avoid dilution effects can be considered as a good strategy to improve fruit composition for premium quality wines (e.g., McCarthy et al., 2000).

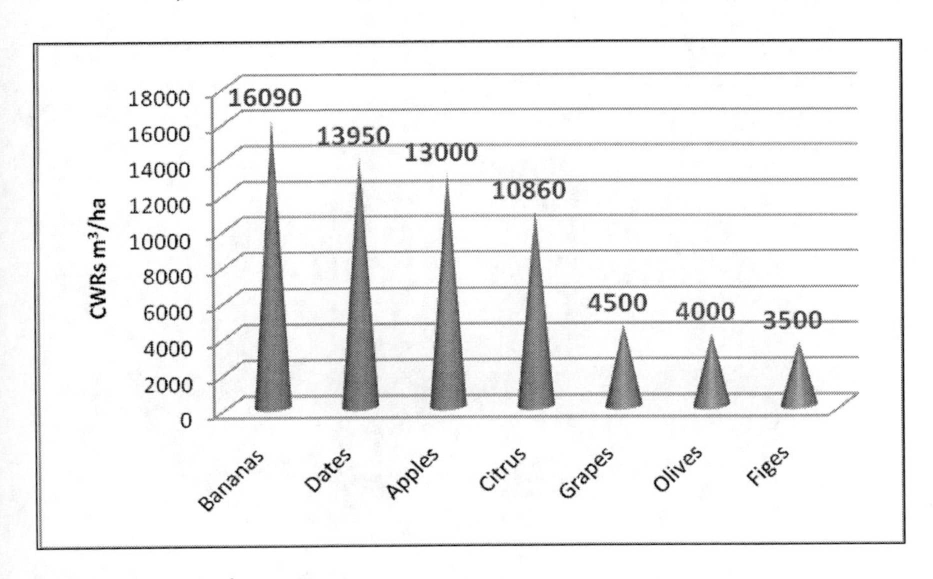

Figure 1. CWRs (m^3/ha) for fruit trees in Jordan.

2. BACKGROUND

2.1. Syria

Syria, 35°00' North latitude and 38°00' East longitude, with a total area of about 185,180 km^2 and a total population of about 20.4 million has a long dry season and a short winter (CBS-SYR, 2010). Depending on humidity and rainfall, Syria can be divided into five zones: wet (annual rainfall about 1000 mm), semi-wet, semiarid, arid, and dry (annual rainfall about 600 mm). The total cultivated area in Syria is about 5,664 thousand ha. Crops include cereals and dry legumes, fruit trees, industrial crops, and vegetables. About 23% of agricultural land is constituted by fruit farming. This figure doubled during the last two decades. Olives occupy more than 60% of the fruit tree land, while grapes occupy only 6% (55.9 thousand ha; Figure 2). Analyzing different fruit tree production between 1998 and 2009 shows that the largest increase was for olive trees from 459.7 to 635.7 thousand ha and for almond trees from 38.2 to 64.2 thousand ha. However, there was a decrease for some fruit trees such as figs from 10.7 to 9.7 thousand ha and quince from 1000 to 500 ha (CBS-SYS, 2010).

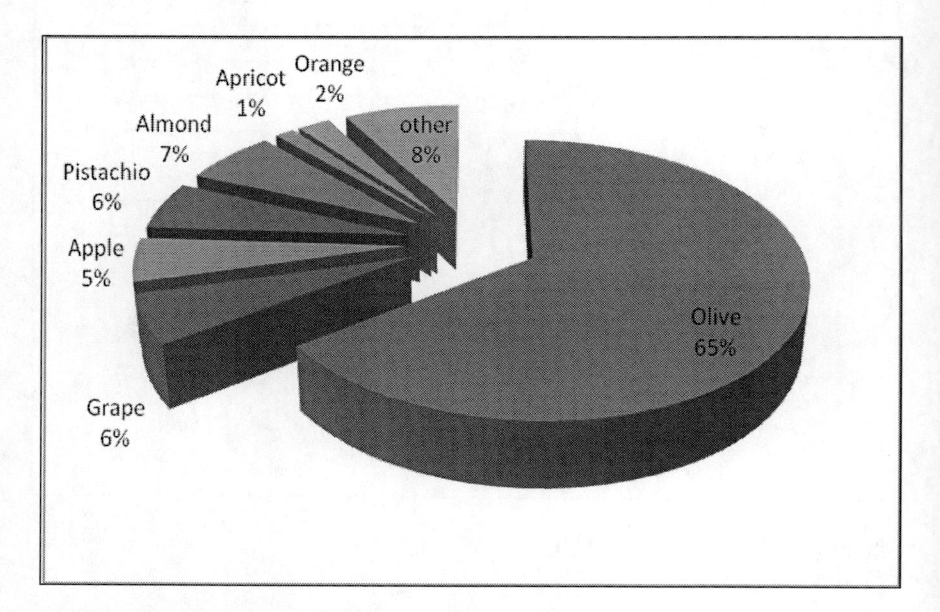

Figure 2. Fruit trees in Syria, 2009.

2.2. Jordan

Jordan with 5.98 million inhabitants on 89,318 km^2 of land, is located between latitude 29.0◦ and 33.5◦ North and longitude 35.0◦ and 39.5◦ East. The country is considered one of the most water scarce countries in the world. The country is characterized by a long, hot, and dry summers and a short winter with an average rainfall range between 40 and 500 mm. The total cultivated lands are about 224,190 ha. 37% of which are fruit trees (DOS-JO, 2010). Grapes occupy about 31.4 thousand ha, 60% of which are irrigated trees that produce about 34475 tons of grapes every year (DOS-JO, 2009).

3. Changing Crop Patterns

There are many factors affecting changing crop pattern, each of which solves the problem from its own horizon. In order to reach water sustainability, however, integrated water resources management (IWRM) is needed. In the following, main factors that should be taken into consideration while planning changing crop pattern are discussed.

1. Virtual Water/Embedded Water

Based on the CWR of grapes, 4500 m^3/ha, and knowing that the total grape cultivated area in Jordan and Syria are 3138 and 55861 ha (DOS-JO, 2009; CBS -SYS, 2009), grape plantation will totally consume about 14 and 252 million cubic meters (MCM) each year, respectively. The embedded water (EW) can be estimated using the following equations:

$$EW= CWR/Y \qquad (1)$$

$$Y= CP/A \qquad (2)$$

where Y is the yield (ton/ha); CP is crop production (ton), and A is crop area (ha). Using the crop area in Eq. (2) will give us the actual embedded water. However, when it comes to the international virtual water flow the total area may be included instead. Hereunder, in Tables 1 and 2, the embedded water in grapes and other fruit trees according to CWRs in Syria and Jordan were estimated. A quick look at these tables shows high EW values for apples and dates in comparison with grapes and pomegranates.

Table 1. Embedded water in some fruit trees in Syria

Fruit tree	Crop area (ha)	Production (ton)	CWR (m^3ha)	EW (m^3/ton)
Olives	444983	885942	4000	2009
Grapes	40779	358000	4500	513
Apples	34943	360978	12192	1180
Dates	129	4037	13950	446
Pomegranates	3482	60055	6000	347

Table 2. Embedded water in some fruit trees in Jordan

Fruit tree	Crop area (ha)	Production (ton)	CWR (m^3ha)	EW (m^3/ton)
Grapes	2456	34475	4500	409
Apples	2261	31111	13000	944
Dates	1235	9681	13950	1780
Pomegranates	218	3490	6000	375

2. Economic Value of Water

An economic analysis was performed regarding some of the above crops. Total cost (TC) and total sale (TS) were estimated according to the local market conditions. The total costs contain: 1) the agricultural operations, which includes tillage, fertilization flatting, irrigation, hoeing and weeding, planting, pesticide control, harvesting, sorting and packaging, and crop transportation; 2) production requirements, which include fertilizers, packages, seeds, water and pesticides; and 3) other costs such as: land rent, capital interest and incidental expenses. The TC was estimated with the help of Ministry of Agriculture and Agricultural Reforms (MAAR) in Syria and Jordan University in Jordan. Figure 3 shows grape cost percentage in Syria in 2009.

Then, gross profit (GP), and gross profit to water use ratio (GP/WU) were estimated by the following equations:

$$GP = TS-TC \quad (3)$$

$$GP/WU= GP/CWR \quad (4)$$

$$GM=GP/TS \quad (5)$$

where: TS is total sales (US$) and TC is total costs (US$). The results of these economic analyses are shown in Tables 3 and 4 for Syria and Jordan, respectively.

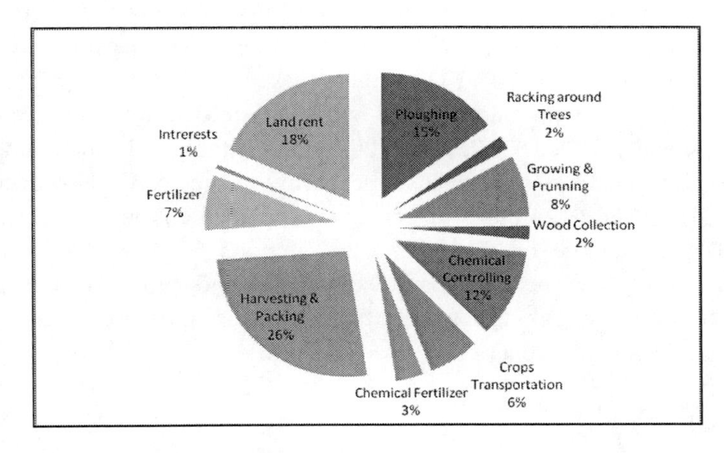

Figure 3. Grape total costs percentage in Syria.

Table 3. Economic analysis for some crops in Syria

Crop	GP (US$/ha)	CWR (m^3/ha)	GP/WU (US$/$m^3$)
Grapes	1690	4500	0.38
Olives	1100	4000	0.28
Citrus	7650	9808	0.78
Apples	3630	12192	0.3
Tomatoes	3130	6916	0.45
Garlic	3500	8777	0.47
Cotton	800	13090	0.06

Table 4. Economic analysis for some crops in Jordan

Crop	GP (US$/ha)	CWR (m^3/ha)	GP/WU (US$/$m^3$)
Grapes	13218	4500	2.9
Citrus	10000	10860	0. 9
Bananas	1087	16090	0.07
Dates	5073	13950	0.36
Hot pepper	1795	3000	0.6
Onion	1380	3000	0.46

3. Country Trade

Changing crop pattern should take the national trade into consideration. Therefore, viewing the national exports and imports is the first step towards changing crop pattern. This will solve the most vital questions; does the country need such a crop? And 1) are there markets for the extra production?

From the national trade data of Syria and Jordan we found that Syria imports annually 109, 107, 25, 10, 500, and 120 thousand tons of tomatoes, garlic, onion, potatoes, soya beans, and barley, respectively. However, Syria exports 31595 and 61243 tons of grapes and combed cotton, which has a low GP/WU ratio. Jordan, on the other hand, imports 5774 and 20000 tons of grapes and apples, respectively, and exports 134 and 20 tons of bananas and apples, respectively. Taking the three factors that are presented above, we can suggest the following scenarios:

For Syria:

As Syria does not import grapes, we have three different scenarios:

- The first scenario depends mainly on changing the exported crops. As it is seen in Table 3 grapes has higher value of GP/WU than cotton. Therefore, Syria may plan to expand grape exports or enhance grape production in Syria. The 61243 tons of cotton, which are exported, need 15580 ha. However this area can become vine farms. In this case, and according to CWRs of grapes and cotton, the saved water from this scenario can be easily estimated by multiplying the difference between cotton's water requirement and grape's water requirement by the cultivated area, which would be 8590 m^3 per ha (13090-4500), which means an annual saving of about 134 MCM.
- In the second scenario, and in order not to affect the industrial sector, we propose to change the cotton land area by other crops that Syria would need. The 15580 ha of exported cotton can be changed into garlic. Accordingly, depending on the CWRs, the saved water from this scenario can be easily estimated as in the first scenario, (13090-8777). The saved water would be 67 MCM.
- In the third scenario, we propose to change the exported cotton land by other crops that Syria would need or export. We estimated the annual amount of saved water from this scenario to be about 106 MCM, Table 5.

Table 5. Saving water by changing crop pattern in Syria

Crop	Grape	Garlic	Onion	Potato	Cotton
Yield (ton/ha)	7.8	7	19.7	20	
Amount (ton)	50000	50000	10000	5000	
Area (ha)	6410	7143	508	250	15580
CWR (m^3/ha)	4500	8777	8777	5641	13090
Consumed water (MCM/year)	28.8	63	4.5	1.4	203.9
Saved water MCM/year					106

Other scenarios would be possible, for example, changing maize and sugar beet areas, which have low GP/WU ratios, by some vegetables that are needed for the national market.

For Jordan:

Because of climatic reasons in Jordan, land used for bananas is not suitable for grape plantation. However, as it has a higher GP/WU, grapes can be planted instead of apples. The total apple land area is about 2307 ha, which needs about 30 MCM of water. However, if this land is planted with grapes, the total consumed water will be about 10 MCM, which means a total saving about 20 MCM per year.

CONCLUSIONS

This chapter tried to assess water value in agriculture for some specific fruit production. The focus in water scarce countries shows a vital need to save water by all means. One of the most effective way in estimating and achieving water saving is the virtual water/embedded water concept. Virtual water together with gross profit to water use ratio should be the basis for any agricultural plan in water-scarce countries such as Jordan and Syria. These countries may reconsider their water and agricultural policies and study them from a virtual water perspective. This chapter shows that grapes have lower embedded water value comparing with some other crops in Syria and Jordan due to its low CWR (4500 m^3/ha). The water embedded in grapes is about 513 and 409 m^3/ton in Syria and Jordan, respectively. Changing crop pattern can help in saving water. For Syria, changing areas that were planted with cotton for exportation in Syria may save about 134 MCM per year. Other scenarios can be proposed for the same target. For Jordan, on the other hand, changing apple land into grapes can save about 20 MCM per year.

REFERENCES

CBS-SYS, Central Bureau of Statistics of Syria. 2010. *http://www.cbssyr.org/* [accessed on 24 April 2011].

DOS-JO, Department of Statistics of Jordan. 2010. Jordan in figures—agriculture surveys. *http://www.dos.gov.jo/agr/agr_e/index.htm [*accessed on 24 April 2011].

Intrigliolo, D. S. and Castel, J. R. 2007. Evaluation of grapevine water status from trunk diameter variations. *Irrigation Science,* 26**,** 49-59.

Intrigliolo, D. S. and Castel, J. R. 2009. Response of grapevine cv. 'Tempranillo' to timing and amount of irrigation: Water relations, vine growth, yield and berry and wine composition. *Irrigation Science,* 28**,** 113-125.

Mourad, K. A., Gaese, H. and Jabarin, A. 2010. Economic Value of Tree Fruit Production in Jordan Valley from a Virtual Water Perspective. *Water Resources Management,* 24**,** 2021-2034.

Santesteban, L. G., Miranda, C. and Royo, J. B. 2011. Regulated deficit irrigation effects on growth, yield, grape quality and individual anthocyanin composition in Vitis vinifera L. cv. 'Tempranillo'. *Agricultural Water Management,* 98**,** 1171-1179.

Shatanawi, M., Nakshabandi, G., Ferdous, A., Shaeban, M. and Rahbeh, M. 1998. Crop water requirement models for crops grown in Jordan. Technical report no. 21. Amman, Jordan: University of Jordan, Water and Environmental Research and tudy Center.

McCarthy, M. G., Loveys, B. R., Dry, P. R. and Stoll, M. 2000. Regulated deficit irrigation and partial root zone drying as irrigation management techniques for grapevines. *Deficit irrigation practices*. FAO Water Reports No 22. Rome, Italy, pp 79–87.

In: Grapes ISBN 978-1-61470-950-3
Editors: R. P. Murphy et al., pp. 193-200 ©2012 Nova Science Publishers, Inc.

Chapter 9

REUTILIZATION OF GRAPE POMACE AS SOLID MEDIUM OF FERMENTATION FOR THE PRODUCTION OF HYDROLYTIC ENZYMES OF INDUSTRIAL INTEREST

A. B. Díaz, I. Caro, I. de Ory, J. Bolívar*[*] *and A. Blandino

Department of Chemical Engineering, Food Technology and Environmental Technologies, Faculty of Sciences, University of Cádiz, 11510 Pto. Real (Cádiz), Spain;

[*]Department of Biochemistry and Molecular Biology, Faculty of Sciences, University of Cádiz, 11510 Pto. Real (Cádiz), Spain

Grape is the most widely cultivated fruit crop in the world. From the world's total production of 60 million tonnes, about 68% of grapes are used for winemaking. Grape pomace is the main residue left after juice extraction from the grapes in the wine making industry and is formed from the skins, seeds and pieces of stem. [1] It constitutes about 16% of the original fruit. The average composition of this medium includes carbohydrates, fibre, fats, proteins and mineral salts. The main component of the fibre is lignin and then hemicelluloses, cellulose and pectin.

Only in Spain, over 250 million kg of grape pomace are generated and re-used as animal feed (with low nutritional value) [2] r nutritive ingredient, [3,4] in the production of citric acid [5] and ethanol by fermentation,[1] in the

[*] E-mail: anabelen.diaz@uca.es; Tel.: 0034956016376

extraction of anthocyanins, [3,6] etc. However, this material has scanty economic profitability, so it is under-exploited and most of it is generally disposed in open areas, leading to serious environmental problems. Thus, the potential utility of this waste for the production of value-added products, such as enzymes by solid state fermentation is promising and it could represent an interesting advantage in the maintenance of the environmental equilibrium and, also, an economic revaluation of the raw material. Moreover, due to 30-40% of the production cost of industrial enzymes is accounted by the substrate of fermentation, the use of low cost substrates is one of the ways to greatly reduce costs.

SSF is defined as the fermentation involving solids in the absence (or near absence) of free water. The substrate must contain sufficient moisture to support the growth and metabolism of the microorganism. This technique has been employed in Japan and Asia from centuries for the production of some fermented food as koji, tempeh or soya sauce, among others. [7] This technique offers the possibility of processing agro-industrial residues that can be used, for instance, as substrates for enzymes production helping in this way to minimize the pollution.

Our research is focused in the revalorization of grape pomace by using it as substrate of fermentation for the production of hydrolytic enzymes such as xylanase, exo-polygalacturonase (exo-PG, a type of pectinase), CMC-ase (a type of cellulase), which are commonly used together to facilitate clarification processes in wine cellars and juice industries. [8] For these studies we growth the filamentous fungus *Aspergillus awamori* on grape residues by using the SSF culture technique. The variety of grape *Palomino Fino*, selected to produce these enzymes by SSF, has a high carbohydrate content (8% in the seeds, 13% in the skin), with the fibre representing about 50% of the total mass. [1] The principal component of fibre is lignin (64% of the fibre in the pips and 59% in the skin), other major components are hemicelluloses (18% of the fibre in pips and 31% in skin) and cellulose (17.75% and 6% in pips and skin, respectively) and the minor component of the fibre is pectin, representing only 0.25% of total fibre in pips and 4% in skin. [1,9] The hydrolytic enzymes obtained degrade polysaccharides of the cellular wall like celluloses, hemicelluloses and pectins, having relevant applications in food industry, for the production of juices and fruit extracts. [10, 11]

The role of pectinases is very important in fruit juices and wine clarification. They enable the increase of the free run juice volume by decreasing its viscosity and improve the clarification of the juice or wine filtration. In association with other lateral activities such as xylanase and cellulose, pectinases speed up the natural process of winemaking and improve the quality of the wine. Although several types of pectinases can be found, polygalacturonases are the most abundant and studied ones, representing around 25% of the industrial enzymes sales. [12] *Aspergillus niger* pectinases are the most widely used in industries because this strain possesses GRAS (Generally Regarded As Safe) status so the metabolites produced by this strain can be safely used. Regarding xylanases, they also increase the degradation of grape skin, enhance aroma and favour the release of phenolic compounds in wines which help to stabilize their colour [13].

In most cases, agroindustrial wastes used to produce these enzymes by SSF or submerged fermentation (SmF) do not contain all the necessary nutrients for this purpose, or maybe they are available in sub-optimal concentrations. In these cases, the substrate must be supplemented to stimulate or improve the enzyme production by adding extra carbon sources or nitrogen sources. [14] Supplementation can also be carried out with the adjustment of the initial moisture content of the residue using a solution containing mineral salts or combining the solid with other residues. [15] In case of grape pomace, we established a protocol for the adjustment of the nutrients composition in order to enhance the production of exo-PG, xylanase and CMC-ase. This process involved the washing of the solid substrate in order to remove excess of reducing sugars, which inhibited the synthesis of the studied enzymes, the supplementation with a solution rich in mineral salts and nitrogen sources and the mix with orange peels in a proportion 1:1 (w/w), which constitute a natural source of pectin, cellulose and hemicellulose (contain 14.4% of pectin and 16.2% and 13.8% of cellulose and hemicellulose, respectively). In these conditions, we detected a maximum xylanase activity of 32.69 ± 3.94 IU/gds after 6 days of fermentation and peaks of pectinase and cellulase activity of 3.77 ± 0.87 IU/gds and 5.35 ± 0.10 IU/gds at the days 4 and 10, respectively. We found that maximums activities for xylanase, exo-PG and CMC-ase were 17, 5 and 16 times higher, respectively, than the ones analysed for grape pomace alone moistened with distilled water and, for xylanase and exo-PG, these maximums were reached earlier. The enzyme activities measured for the mixture of orange peels and grape pomace are of the same order of magnitude (or even higher in some cases), to those produced by using other agro-industrial wastes in SSF. [16]

We have also established an adequate protocol for the concentration of the enzyme extracts obtained from grape pomace fermentation, in order to reduce the volume and facilitate their applicability on different processes [17] With those purposes, the methods of precipitation and lyophilisation are commonly reported in literature. Protein precipitation is produced by adding salts, such as ammonium sulphate, or organic solvents (acetone, ethanol, etc.) which decrease the solubility of the proteins. [18,19,20] Lyophilisation consists on water elimination of a substance with its previous freezing, reducing the pressure to allow its sublimation directly from the solid to the gas phase. With this technique, it's possible to dry a sample at low temperatures without damaging it.

Several methods were used to concentrate the hydrolytic enzymes produced by SSF on grape pomace: precipitation by adding ethanol, acetone or a saturated ammonium sulphate solution at two temperatures (4° and −20 °C) and lyophilisation for 24 h. From them, ethanol precipitation at 20ºC or 4ºC was the best method to concentrate exo-PG, with an efficiency of 50.23% (calculated as the percentage of the concentrated extract activity related to the theoretical one). Ethanol was more efficient than acetone for exo-PG concentration, however, for xylanase, the best results by precipitation were measured with acetone at 4ºC, obtaining an efficiency of the process of 49.26%. [21] Lyophilisation was also a good method to concentrate exo-PG and xylanase, however it shows the disadvantage of being an expensive technique, which needs a long drying time.

We have used the concentrated protein extract obtained after the crude enzyme precipitation with ethanol at −20 °C for grape must clarification studies. In order to evaluate its efficiency, it was compared with commercial crude enzymes used in wine industry: Enovin Clar (Agrovin) and Enozym Vintage (Agrovin). [21] Both are enzymatic preparations, with no defined specific activities, which are used in wine-cellars as an aid for clarification by sedimentation. The specifications of both preparations show that they contain high pectinase activity and recommend the addition of 10 mg of solid per 500 mL of must. For must clarification experiments, 450 mL of white must of the Palomino Fino variety were added to 500 mL volumetric cylinders. 10 mg of commercial crude enzymes or the same amount of the concentrated extract obtained in this work were dissolved in 50 mL of must and poured into the volumetric cylinder. The mixture was agitated thoroughly and incubated at 25°C in static conditions to let the must decantation. After 19 and 40 h of incubation, samples were taken from the middle of the volumetric cylinder and turbidity was determined with a turbidimeter, expressing the results in NTU (Nephelometric Turbidity Units). A sample which only contained must was used as control.

When this concentrated protein extract was used for clarification of must, it decreased a 97.5% the initial turbidimetry after 40 h of incubation. In this case, the grade of clarification was higher than the one obtained with the commercial crude enzymes Enovin Clar and Enozym Vintage, which turbidimetries were 3.1 and 2.1% higher, respectively than the one analysed with the concentrated protein extract obtained from grape pomace fermentation. These results could be explained with the presence of other enzymes, such as xylanases and cellulases, measured in the concentrated extract obtained by SSF on grape pomace, which could help pectinases in the clarification process. Alternatively, a higher concentration of the enzymes responsible of the clarification in the concentrated extract compared with the commercial protein extracts could also explain the difference. In spite of the commercial crude enzymes also contain different enzymes (esterases, polygalacturonases, pectin lyases, etc), their exact nature and ratios are not specified, which may be crucial in obtaining optimal performance in wine making [22] It has been described that the addition of a correct balance of exogenous enzymes to complement the poor endogenous enzyme activities of the grape is a key factor to improve must clarification. [23] For example, the use of a macerating enzyme preparation having a blend of activities (pectinases, cellulases, and hemicellulases) from *Trichoderma* and *Aspergillus* improved the clarification process of must. [24].

Stability of enzymes is another important factor to be considered for its efficient industrial application. Enhancing the stability and maintaining the desired level of activity over a long period are two important points to consider for the selection and design of enzymes. [25] For these reason we also studied the stability of exo-PG and xylanase in extracts obtained after SSF on grape pomace in the presence of several cations, in a pH range from 3.0 to 11.0 and at different incubation temperatures from 4 to 70 °C. Results revealed a high stabilibity of the enzymes, which retained practically 100% of their activity in the range of the pH evaluated, and in presence of different cations, only showing an important inhibition with Cu^{2+} and Co^{2+}, besides Mn^{2+} in the case of xylanase. With regard to the thermal stability, the enzymes were very stable at 4 °C and 30 °C, mainly xylanase which retained practically full activity in those conditions. Even at 45 °C and 50 °C, half-life times of 2.75 and 10.5 h were measured for exo-PG and xylanase, respectively. However, a rapid loss of activity was observed at 70 °C for both enzymes, in which they showed half-life times for deactivation of 2.05 and 7.21 min, respectively [21]

Since some industrial processes require purified enzymes, we have proposed a protocol for partial purification of xylanase and exo-PG by anion exchange chromatography on DEAE-cellulose (dietil aminoetil cellulose) column [21] Besides separating both enzymes in different elution fractions, they were enriched after chromatography, obtaining for exo-PG and xylanase fractions with specific activities 35.5 and 5.51 times higher, respectively than the ones analysed in the culture supernatant.

In conclusion, the reutilization of grape pomace as solid substrate of fermentation to produce hydrolytic enzymes constitutes a new process, with low environmental impact and costs, to revalorize this agro-industrial residue. Exo-PG, xylanase and CMCase activities produced on mixtures of grape pomace and orange peels were similar, or even higher, to those produced by using other agro-industrial wastes, demonstrating therefore its potential utility as an alternative substrate for the production of enzymatic extracts. The extracts obtained reveal a high stability of the enzymes in a wide pH range, at low and moderately high temperatures and in the presence of different cations. In addition, when they were applied for the clarification of grape must, higher yields than the ones achieved with commercial crude enzymes were obtained. All these results suggest the promising application of the enzyme extract produced by the proposed process in juice and wine industries.

REFERENCES

[1] C. Botella, I. de Ory, C. Webb, D. Cantero, and A. Blandino. Hydrolytic enzyme production by Aspergillus awamori on grape pomace. Biochemical Engineering Journal, 26 (2005), 100-106.

[2] A. Larwence. Feed value of grape marc. VI. Extraction, fractionation, and quantification of condensed tannins. *Chemical Abstracts,* 116, (1991), 20017c.

[3] Y. Lu and L. Yeap. The polyphenol constituents of grape pomace. *Food Chemistry*, 65 (1999), 1-8.

[4] A. Ana, N. Croitor, B. Segal, I. Sas. Bioactive substance concentrate from grape marc. *Chemical Abstracts,* 124 (1995), 28617j.

[5] Y. Hang. Citric acid and its manufacture with *Aspergillus niger* from grape pomace. US Patent, US 4791058; *Chemical Abstracts,* 110 (1988), 93581t.

[6] F. Francis. A new group of food colorants. *Trends in Food Science and Technology*, 3 (1992), 27–31.

[7] A. Pandey A, C. Soccol, J. Rodriguez-Leon and P. Nigon. Solid State Fermentation in Biotechnology: Fundamentals and Application. *Bioresource Technology*, 82 (2002), 305.

[8] C. Botella, A. Diaz, I. de Ory, C. Webb and A. Blandino. Xylanase and pectinase production by *Aspergillus awamori* on grape pomace in solid state fermentation. *Process Biochemistry*, 42 (2007), 98-101.

[9] P. Blanco, C. Sieiro and T. Villa. Production of pectic enzymes in yeasts. *FEMS Microbiology Letters*, 175 (1999), 1-9.

[10] E. Rodrigues, A. Milagres and A. Pessoa. Xylanase recovery: effect of extraction conditions on the AOT-reversed micellar systems using experimental design. *Process Biochemistry*, 34 (1999), 121–125.

[11] L. Castilho, T. Alves and R. Medronho. Recovery of pectolytic enzymes produced by solid state culture of *Aspergillus Niger*. *Process Biochemistry*, 34 (1999), 181–186.

[12] R. Jayani, S. Saxena, R. Gupta. Microbial pectinplytic enzymes: a review. *Process Biochemistry*, 40 (2005), 2931-2944.

[13] H. Rehm and G. Reed. Enzymes, biomass, food and feed. *InBiotechnology*, 9 (1996). Weinheim, Germany: VCH.

[14] S. Patil, A and Dayanand. Production of pectinase from deseeded sunflower head by *Aspergillus niger* in submerged and solid-state conditions. *Bioresource Technogy,* 97 (2006), 2054-2058.

[15] [1] V. Papinutti, F. Forchiassin. Lignocellulolytic enzymes from *Fomes sclerodermeus* growing in solid-state fermentation. *Journal of Food Engineering,* 81 (2007), 54-59.

[16] A. Díaz, I. de Ory, I. Caro and A. Blandino. Enhance hydrolytic enzymes production by Aspergillus awamori on supplemented grape pomace. *Food and Bioproducts Processing*, (2010), *in press*.

[17] A. Díaz, J. Bolívar, I. de Melo, I. Caro, I. de Ory and A. Blandino. Comparison of different methods of enzyme concentration for the obtention of enzymatic extracts with high pectinase activity. 18th International congress of chemical and process engineering. *Summaries 5* (2008), 1989.

[18] N. Wang. Enzyme purification by acetone precipitation biochemical engineering laboratory. Department of Chemical & Biomolecular Engineering. University of Maryland.

[19] A. Farag and A. Hassan. Purification, characterization and immobilization of keratinase from *Aspergillus oryzae*. *Enzyme and Microbial Technology*, 34 (2004), 2, 85-93.

[20] H. Fernández-Lahore, E. Fraile and O. Cascone. Acid protease recovery from a solid-state fermentation system. *Journal of Biotechnology,* 62 (1998), 83-93.

[21] A. Díaz, J. Bolívar, I. de Ory, I. Caro and A. Blandino. Applicability of enzymatic extracts obtained by solid state fermentation on grape pomace and orange peels mixtures in must clarification. *LWT - Food Science and Technology,* 44 (2011), 840e846.

[22] G. Harman and C. Kubicek. Trichoderma and Gliocladium. In Enzymes, biological control and commercial applications. London, UK (1998).

[23] A. Roldán, V. Palacios, X. Peñate, T. Benitez and L. Perez. Use of Trichoderma enzymatic extracts on vinification of Palomino fino grapes in the sherry region. *Journal of Food Engineering*, 75 (2006), 375-382.

[24] H. Tsujibo, T. Sakamoto, N. Nishino, T. Hasegawa and Y. Inamori. Purification and properties of three types of xylanases produced by an alkalophilic actinomycete. *Journal of Applied Bacteriology*, 69 (1990), 398-405.

[25] N. Sathyanarayana and T. Gummadi. Purification and biochemical properties of microbial pectinases-a review. *Process Biochemistry,* 38 (2003), 7, 987-996.

In: Grapes ISBN 978-1-61470-950-3
Editors: R. P. Murphy et al., pp. 201-247

Chapter 10

GRAPES: CULTIVATION, VARIETIES AND NUTRITIONAL USES

P. D. Gurak[1], I. Santana[1], A. P. Gil[1], S. P. Freitas[1], M. H. Rocha-Leão[1] and L. Cabral[2]
[1]Federal University of Rio de Janeiro
[2]Embrapa Food Technology

ABSTRACT

Grape is one of the oldest cultivated crops in the world. The genus *Vitis* includes two subgenera: *Euvitis,* or true grape, and *Muscadinia. Vitis vinifera* the only species of European origin, and *Vitis labrusca* (Native American grapes), both *Euvitis*, presents several varieties including both black and green or pale grapes. Around 75 thousand Km² of the land world are intended to grapes cultivation and over 90 % of the production is related to *V. vinifera, V. labrusca,* and its hybrids. More than 70 % of world grape production is destined for winemaking, the remainder being employed in the manufacturing of juices, jams, vinegar, raisins and for *in natura* consumption. Grapes are non-climacteric fruits, botanically classified as berries and its chemical composition plays a key role in the quality criterion.

A variety of factors such as cultivar, climate conditions, soil, vineyard management and maturity influence over macro and micronutrients and bioactive compounds composition. Despite different compositions inherent to each variety, water and carbohydrates (mainly glucose and fructose) are the major macro components and potassium, the most important quantitative mineral. The functional aspect of grapes and grape products given by the bioactive substances brought about a diversity of studies concerning this fruit, since their strong antioxidant capacity is related to health benefits, especially over diseases linked to oxidative stress and inflammation, like cardiovascular and neurodegenerative illnesses and cancer.

The main phenolic compounds present in the fruits that exert such protection are the phenolic acids (derived from cinnamic acid and benzoic acid) and stilbenes (resveratrol), from the non-flavonoids group, and the following flavonoids: flavanols (catechin, epicatechin and epigallocatechin), flavonols (kampferol, quercetin and myricetin), tannins and anthocyanins, the later only present in red varieties. Thus, the beneficial effects will be dependent on the variety of the grape and the kind of consumed product, where red wine and red grape juice comprise the most significant sources, presenting a high marketable value. Residues from grape processing represent an imminent environmental risk due to the high generated volume in a short time and their polluting characteristics.

However, some residues have been highlighted for reuse for its interesting composition, overcoming environmental issues and adding value for these by-products. This review will discuss about the cultivation of grapes, the main forms of consumption and its nutritional and functional implications in human health.

1. Introduction

Grapes and grape products are known worldwide and sustain a market with high added value, which uphold the demand for research in the fields of agronomy, technology and nutrition. This chapter focuses on giving a comprehensive view over the botanical and cultivation aspects of grapes; which varieties are suitable for technological application and the main consumed products; to clarify the influence of the bioactive compounds in the metabolic pathways that provides the nutritional benefits of grapes; and to update information on the important issue of waste generated in the grape industry.

2. Varieties and Cultivation

2.1. The History of Grapevine

Grape is unique: it is a major global horticulture crop and its history is as old as the humankind. Grapes were, in fact, one of the earliest cultivated fruits on Earth, influencing different cultures. In Mediterranean cultures, for example, wine has always had a major role in the life of citizens, who had considered once that "the wine sprang from the blood of humans who had fought the gods". In other cultures, the wine was considered divine, a drink of the gods: Dionysus and Bacchus, according to literature, were dedicated to this beverage, for example [1, 2, 3, 4]. In addition, the historical evidence of vine cultivation and wine throughout the Old World is plentiful and its supporting archaeological evidence has been increasing [3, 5, 6, 7, 8, 9; 10].

Grapes are of the *Vitaceae* family; it is the *Vitis* genus that is of major agronomic importance. It consists of around 60 inter-fertile species distributed throughout Asia, North America and Europe under subtropical, Mediterranean and continental – temperate climatic conditions. Among them, *Vitis vinifera* is the species extensively used in the global wine industry. It is also the only species of the genus indigenous to Eurasia and is suggested to have first appeared around 65 million years ago. Two forms still co-exist in Eurasia and in North Africa: the cultivated form, *Vitis vinifera* subsp. *vinifera* (or *sativa*) and the wild form *Vitis vinifera* subsp. *silvestris* (or *sylvestris*), which is sometimes referred to as a separate subspecies. This historical separation into subspecies was based on morphological differences [2, 3, 11, 12].

Thousands of *Vitis vinifera* cultivars exist but the global market for wine production is dominated by only a few cultivars owing, to a great extent, to how wine is currently marketed. By contrast, the wild form is rare, extending from Portugal to Turkmenistan, and from Rhine riverside to the northern forests of Tunisia. It is believed to be the ancestor of present cultivars and is still observed growing on the canopy of surrounding trees.

The domestication of grape seems to be linked to the discovery of wine, being the earliest evidence of its production found in Iran at the Hajji Firuz Tepe site in the northern Zagros Mountains circa 7400-7000 BP. Seeds of domesticated grapes dated from around 8000 BP were also found in Georgia and in Turkey. Nevertheless, remains of seeds discovered in the Neolithic period in Western Europe and of wild seeds dating of the bronze-age also suggest exploitation of grape at these times [3, 11, 13, 14].

From the primo-domestication sites, there was gradual spread to adjacent regions such as Egypt and Lower Mesopotamia (circa 5500–5000 BP) with further dispersal around the Mediterranean, following the main civilizations (Assyrians, Phoenicians, Greeks, Romans, Etruscans, Carthaginians). Grape cultivation reached also China (2nd century) and Japan (3200 BP) and under the influence of the Romans, *Vitis vinifera* expanded inland, reaching many temperate regions of Europe, as far north as Germany. The Romans were, in fact, the first to give names to cultivars but it is difficult to relate them to modern grapes. By the end of the Roman Empire, however, grape growing was common in most of the European locations where they are grown today [2, 3].

The expansion of vine followed the main trade routes (i.e. rivers such as the Rhine, Rhone, Danube and the Garonne) and an important role for the spread of grape (particularly table grapes) is, as well, the extension of Islam to North Africa, Spain and Middle East [2, 3].

It is believed that during domestication, the biology of grapes underwent several dramatic changes to ensure greater sugar content for better fermentation, greater yield and more regular production, presenting, as a consequence, changes in the size berry and seed morphology. At the time of the extension of Islam to North Africa, the differentiation of table and wine grape was probably already in place in addition to the different color types [2,3].

Following the Renaissance (16th century), *Vitis vinifera* were colonized in new regions (New World countries) where it was not indigenous. The missionaries introduced it to America, first as seeds (because they were easy to transport) and then by cuttings from their places of origin. Cuttings were also introduced to South Africa, Australia and New Zealand in the 19th century, and later to North Africa [2].

At the end of the 19th century, after several millennia of geographical expansion, disease-causing agents from America reached Europe (mildews, *Phylloxera*), resulting in devastation and destruction of many European vineyards, drastically changing the diversity of this species.

As a result, a reduction of the diversity occurred most likely in both cultivated and wild grapes. The "*Phylloxera* crisis" that affected European vineyards had a considerable impact on both cultivated and wild varieties. As a result, modern wild grapevines were endangered and threatened with extinction [2, 15].

2.2. Agronomic Information of Grapevine

The grapevine is a perennial woody plant that can live with proper care for many years, generally for several decades of productive growth. The leaves, steams, and roots are biologically classified as a higher form plant. Such plants are able to use sunlight to make sugar from carbon dioxide and water through the process of photosynthesis. While some of these sugars are used in the metabolic processes of the plant's life system, portions of these precious sugar compounds are stored in the grapevine's roots until *veraison,* a French term referring to a time when grapes commence to ripen and sugars are transported from roots to berries [16].

Considering weather conditions, grapes are, generally, native to temperate zones, primarily between 40 and 50° south latitude. The grapevine can, however, adapt to some rather difficult climate and soil conditions. In reproduction, vines often form *clones*, which are offspring with slight variations in character responding to natural selection processes: while some clones may be more resistant to cold winter temperatures, others may require a shorter growing season, or produce heavier crops, and so forth [16].

Taxonomically, grapevine is classified as it follows: group Spermatophyta, division Tracheophyta, subdivision Pteropsida, class Angiosperma, subclass Dicotyledoneae, order Ramnales, family Vitaceae (Ampelidaceae), genus *Vitis*, subgenera *Euvitis* and *Muscadinia* [16].

Grape is the simplest fruit of hypogynous and the one that confirms exactly the botanical definition of a berry. Remnants of floral parts other than the ovary are absent or vestigial and the developed ovary tissue is fleshy, succulent, and homogeneous. The number of berries per cluster, however, is greatly influenced by: the health and the vigor of the vine, the environmental conditions during flower blooming, among other factors. Bunches of grapes may be composed of less than 10 berries each, for some of the more obscure varieties, and more than 300 berries, for some of the most prolific types. Cluster shapes range from long, cylindrical forms (with or without shoulders), to shorter conical structures.

Generally, each berry consists of a pericarp and seeds, being the pericarp divided into exocarp (skin), mesocarp (flesh) and endocarp. The skin is composed of the epidermis (thickness 6.5-10 μm), which has smaller cells than other dermal layers and the hypodermis (thickness 107-246 μm) with some flesh cells. Mesocarp tissue consists of a pulp of about 25-30 layers of cells, where most of the constituents are stored during ripening. The skin cells have distinct and active metabolism involving many physiological and biochemical changes, occurring during development and ripening with its portion accounting for 5-12 % of the total berry weight. The cells of the pulp have large vacuoles containing the cell sap, which is the main constituent of the juice found in the berry at maturity and accounts for 64-90 % of the berry weight. The seeds constitute up to 10 % of the weight of the fruit [16, 17, 18]. In addition, the grapes are non-climacteric fruits that are edible at the time of picking and have no postharvest ripening cycle [18].

Chemical composition is one of the most important quality criteria for fruit products. The chemical composition of grape is influenced by a variety of factors such as climate conditions (the maximum, minimum, and average temperatures; the rainfall and solar energy level), soil (must be moderate), cultivar, vineyard management (pruning and training systems, fertilization, irrigation, application of growth regulators, and pest control measures), and maturity. Each of these factors exerts its own influence, but complex interactions among these factors also influence the chemical composition of grape.

Generally, the size and composition of the berries are a role in seed development: the greater the number of seeds, the heavier the berry, with relatively lower sugar and nitrogen concentration, but higher of acidity [19]. In addition, Morris [20] has shown that mechanically harvested grapes can be of a better quality than hand-harvested ones. Effects on the quality of machine-harvested grapes can be altered or influenced by the type of machine, the cultivar, the system production, the harvest temperature and the interval between harvesting and processing, and the postharvest handling system.

2.3. Grape Species and Varieties

Selection of the vine varieties and cultivars is obviously fundamental to the product line. Such as wine, grapes are grown for the production of juice, wine, and others products, as well as to be sold as fresh fruit. There are perhaps 5000 named varieties of *Vitis vinifera* and 2000 of *Vitis labrusca* and other species. In addition, there are a very large number of hybrids between *Vitis vinifera* and American species – the so-called “direct producers”. Among these varieties, grapes may be white, green, pink, red, or purple in color when ripen. They may, as well, have small or large fruit and clusters and may ripen very early to very late. The shape of their berries, clusters, and leaves vary and the texture may be pulpy and solid or soft and liquid [21].

Grapevine belongs to the genus *Vitis* of the family Ampelidaceae (Vitaceae), which includes several species. They are classified into four groups according to their origin: 1) *Vitis vinifera* or European grape, which is subdivided into many varieties, bearing black or green grapes; 2) American vines (such as *V. riparia, V. ruperis, V. labrusca*), generally bearing black grapes; 3) French hybrids and *V. rotundifolia* or muscadine, which is tolerant to hot conditions; and 4) Asian vines (*V. amurensis*) which are not of economic importance. Many vines either do not bear fruits or produce grapes unsuitable for human consumption or for winemaking. However, the crossing of American species with *V. vinifera* has resulted in the production of hybrids with good cultural qualities such as low production costs.

There appear to be four kinds of reasons to select a certain grape variety: tractability, distinctive flavor, other special characteristics, and economical-sales. These are not necessarily mutually exclusive and compromises may be required. The ultimate in tractability are those grapevines that are relatively easy to grow and very productive. Other special characteristics important to the selection of which varieties to grow include factors as length of ripening and the usual harvest date, the retention of acidity in harvest, or the tendency regarding certain processing problems as product derivate of grapes are made. Which variety should be planted in a specific new vineyard depends on many factors as biology, climatology, and economics. Not a single variety can be recommended universally, nor should it be, considering the value of diversity among products derivate of grapes to maintain consumer interest [18].

Each species of grapevine is identified by genetic characteristics that influence its vegetative, productive and qualitative behavior, factors that are more or less stable in various environments. Many parameters are determined mostly by the genetic patrimony, particularly the trends of phenological phases or the basic characteristics of the cluster and the must (e.g., phenolic and aromatic quality and richness, sugar/acid ratios). However, different varieties have different reactivity and stability with respect to the cultivation site. In fact, there are vine varieties that are well adapted even in dissimilar environments, always guaranteeing satisfying results (e.g., *Cabernet Sauvignon* and *Chardonnay*, grown worldwide), and that interact minimally with the surrounding environment, compared to vine species that, on the contrary, readily respond even to small pedological and climatic variations, such as *Sangiovese* [22].

The current distribution of *Vitis* species includes northern South, Central and North America, Asia and Europe. In contrast, species in the subgenus *Muscadinia* are restricted to the southeastern of The United States and the northeastern Mexico[23]. More than 8000 grape cultivars are known: all of them able to be fermented into a type of wine when crushed, and most of them able to be dried or eaten fresh. Similarly, the raisins of commercial interest are produced mainly from three cultivars, namely, *Thompson Seedless*, *Black Corinth*, and *Muscat* of *Alexandria*, and less than a dozen cultivars are grown extensively as table grapes. Most of the sweet juice comes from *Concord*: and only one or two seedless cultivars (*Thompson Seedless* and *Canner*) are used for canning in combination with other fruits in fruit salad, and fruit cocktail.

2.3.1. Vitis Vinifera Grapes

Hundreds of varieties of *Vitis vinifera* are cultivated around the world today as the “old World” vine. More than 90 % of the wine grown globally is from this genus and species. Their productivity is usually lower, and their cultivation is typically more demanding. Their reputation usually comes from their yielding fine, variedly distinctive, balanced wines with long-aging potential. In favorable locations, their excellent winemaking properties compensate the reduced yield and the increased production cost [16,23].

The major Vitis vinifera White Varieties: Chardonnay, Sauvignon Blanc, Johannisberg Riesling, Semillon, Gewurztraminer, Muscat de Frontignan, Pinot Blanc, Sylvaner, Pinot Grigio, Chenin blanc, Ehrenfelser, Müller-Thurgau, Muscat blanc, Parellada, Semillon, Traminer, Viognier, Viura. The major Vitis vinifera red varieties: Cabernet Sauvignon, Merlot, Pinot Noir, Gamay, Sangiovese, Nebbiolo, Zinfandel, Syrah, Cabernet Franc, Mourvedre, Petite Sirah, Barbera, Dolcetto, Gamay, Grenache, Graciano, Tempranillo, Zinfandel.

2.3.2. Vitis Labrusca Grapes

This species is the so-called “fox” grape. Some people contend that the character of *labrusca* varieties is related to once or another of the odors associated with foxes. More plausible legends relate to foxes being attracted by the intense fruit flavor – often associated with “grape” soft drinks, candies, and so forth [16].

American hybrids constitute the major plantings in eastern North America, and they are grown commercially in South America, Asia, and Eastern Europe. Among the American hybrid cultivars, the most important are based on *Vitis labrusca*. They possess a wide range of flavors, such as “*Niagara*” which are characterized by the presence of a foxy aspect. Others, however, are characterized more by a strawberry fragrance (“*Ives*”), strong floral aroma (“*Catawaba*”), among others. The major *Vitis labrusca* White varieties:

Catawba, Delaware, Niagara; and the major *Vitis labrusca*Red varieties: *Concord, Ives* [23].

2.3.3. Vitis Riparia

This species is referred to as the "post-oak" grape by some viticulturists. It is widely adapted to almost all temperature of North America, but is mostly found in the east of the Rocky Mountains, from Canada to the Gulf Coast. Because of its hardiness in cold-weather climates, it is often used for grafting as a rootstock or in grape breeding by researchers [16].

2.3.4. Vitis Rotundifolia

The subgenus *Muscadinia* is more commonly known in America's Deep South as *Muscadine*. It has generally small, loose clusters of large berries typified by a thick skin and dense pulp consistency and literally bursting with jammay-fruit flavors. The bronze-skinned varieties are collectively referred to as Scuppernongs and the reds as Muscadines. Most are, actually, hybrid cultivars in the *Vitis rotundifolia* species, with the most commercially important being *Carlos*, *Magnolia*, *Noble*, and *Tarheel* [16].

2.3.5. Hybris and Grafting

Natural variations in grapevine and/or fruit morphology result in what are termed *varieties*. Some varieties have certain mutations apart from a morphological change that is differentiated by the term *clone*. Varieties and clones are both the result of natural selection [16].

Hybrids are, however, products of the human interest in grapevine breeding. The process is one whereby pollen is gathered from the male stamens of selected grape flowers and used to fertilize female pistils in other specific grape flowers. In the resulting meiosis, a genealogical combination takes place in seeds that reflects the attributes of each parent. Each new vine resulting from hybridization is called a *cultivar* – term derived from "*cultivated variety*" [16].

The rationale for hybridization is to improve resistance to disease and adverse climate/soil conditions to increase yields and advance fruit quality. Unfortunately, it is common to find negative characteristics associated with dominant genes. History indicates that it may take hundreds of crosses to find just one cultivar worthy of commercial cultivation. University breeding programs continue, but state-of-the-art research looks to the advent of biotechnology, which may serve to make classic varieties increasingly resistant to adverse climates and disease, with greater productivity of better quality fruit [16].

Hybridizing should not be confused with *grafting*. In avoiding negative soil conditions, viticulturist will often graft one variety upon another. Great care must be taken to ensure that the *xylem* and *phloem* sap-flow tissue matches perfectly between the *scion* (upper fruit-bearing variety) and the *rootstock* (root-support variety). Vines can be "bench-grafted" and propagated asexually in a nursery or "to-grafted" to change a mature vine from one variety to another in a vineyard [16].

2.3.6. French-American Hybrids

The term *French-American Hydrid* evolved from efforts to find vines that were resistant to the great *Phylloxera* root louse blight that ravaged most of the world wine-growing community during the latter 19th century. French *vinifera* vines, well known for their fine wine quality, were bred with disease-resistant *labrusca* and *riparia* American vines [16, 23]. The major French-american Hybrid White cultivars are: *Seyval Blanc, Vidal Blanc, Vignoles, Dutchess;* and the major French-american Hybrid Red cultivars are: *Chambourcin, Chancellor, Marechal Foch, Delaware*and *Noble.*

Finally, varieties of seedless grapes are of relatively recent origin and have been derived from crosses between American-seeded and seedless *Vitis vinifera* varieties. The best-known varieties are Interlaken Seedless, *Himrod*, *Canadice*, *Lakemont*, *Suffolk Red*, *Concord Seedless*, *Glenora* and *Romulus*.

3. Main Forms of Consumption

According to the Food and Agriculture Organization (FAO), 75,866 square kilometers of the world are dedicated to grapes. Approximately 71 % of the world grape production is used for wine, 27 % for fresh fruit, and 2 % for dried fruit. A portion of grape production goes to producing grape juice to be reconstituted for fruits canned "with no added sugar" and "100 % natural". The area dedicated to vineyards is increasing by about 2 % per year [24].

The use widely varies from one country to another, often depending on the physical and politico-religious (wine prohibition) conditions of the region (Jackson, 2000). Indeed, the volume of grapes processed for beverage used worldwide (including wine and beverage alcohol) does not exceed any other fruit.

3.1. Wines

Wine has been the most extensively consumed for pleasure beverage since thousands of years. Grapes and their juices can be converted into wines of different grades and quality depending upon the craftsmanship and the maintenance of the distinctive quality of wines from different regions. The manufacture of wine forms the major industrial use of the grape involving complex biochemical processes in transformation of grape juice into red, white, blush or rosé wine [4,23]. Wine styles often reflect the unique climate and political-economic environment from which they emerged, what might influence the development of styles such as noble rot and sparkling wines. Many authors, such as Peynaud & Ribereau-Gayon19, Amerine, Bergs & Cruess [21]; Boulton et al. [18]; Jackson [23], among others, have detailed the production operations of some wines.

3.1.1. Wine Classifications

Sweetness, alcohol content, carbon dioxide content, color, grape variety, fermentation or maturation processes, or geographic origin may group wines. However, for taxonomic purposes, they are often divided into three categories: still, sparkling, and fortified wines, what demonstrates significant differences not only in the production but also in the use.

Moreover, wines are commonly subdivided according to the geographic origin. In many European countries, this is associated with the traditional use of particular grape cultivars, as well as grape-growing and winemaking techniques. In the New World, wines may be classified similarly, but few regions have become consistently associated with particular styles.

Because geographic wine classifications are based on the sensory characteristics of the wine, the arrangement here is based primarily on stylistic differences. Wines are initially grouped based on alcohol concentration, what is commonly indicated by the terms “table” (alcohol contents raging between 9 and 14 % by volume) and “fortified” (alcohol contents ranging between 17 and 22 % by volume), being the first ones subdivided into “still” and “sparkling” categories, depending on the wine’s carbon dioxide content [16, 23].

Sparkling wines are “effervescent” or bubbling, being often classified by their method of productions. The main techniques are “champenoise” (the traditional champagne), and bulk (charmat) methods. Nowadays, there are people who still use the terms Champagne and sparkling wine interchangeably, but this is considered incorrect. Champagne is a sparkling wine produced in France, being thus a generic term. In other words, all

Champagnes are sparkling wines, but not all sparkling wines are Champagne. The sparkling wine fermentation process, which is taken in tanks, is called Charmat or the bulk [16, 23].

The driest sparkling wines are generally referred to as *naturel,* while very dry is *brut* or *sec* – these being the most popular. *Demi-sec* is noticeably sweet and *doux* very sweet. One finds *dry* and extra *dry* on sparkling wine labels, too, and these are often between sec and demi-sec in sweetness level [16].

Carbonated sparkling wines (deriving their sparkle from carbon dioxide incorporated under pressure) show an even wider range of styles. These include dry white wines, such as vinho verde (historically obtaining its sparkle from malolactic fermentation); sweet sparkling red wines, such as *Lambrusco*; most crackling rosés; and fruit-flavored "coolers" [23].

Dessert wines (or fortified wines or/and Appetizer wines) are produced in a wide range of styles, usually made by the addition of grape brandy to a fermenting juice or must, and less often by completely fermented table wines. The brandy addition usually increases alcohol content up to 19 to 20 % by volume (not to be exceeded in 24 %), such as fine-style Sherries and dry vermouths – the second ones being flavored with a variety of herbs and species. More commonly, fortified wines possess a sweet character, such as oloroso sherries, ports, madeiras, and marsalas [16, 23].

Botrytized are wines unintentionally made from grapes infected by *Botrytis cinera,* which have undoubtedly been made for centuries. The fungus is omnipresent and produces a destructive bunch rot under any climatic conditions. Wines made from bunch-rotten grapes are unpalatable. Nevertheless, under unique climatic conditions, infected grapes develop what is called noble rot. These grapes beget the most seraphic of all white wines [23].

Pop wine is a type of wine thought by some to be named after its popularity among young adults and ethnic groups. Others will testify that pop wines have a similarity to soda pop. Perhaps both sources apply. Commercially produced, pop wines may be made at a alcohol levels under 14 % by volume, closely resembling aperitif wines, except that the added essences are more exotic, typically from pronounced and/or berry flavors [16].

3.2. Grape Juice

Grapes are processed into nonalcoholic grape juice, which is an important commercial product. Whole grapes of a more selected cultivar are utilized to get a proper ratio of sugar to acidity. There are several options for juice extraction and subsequent treatment. The hot break process is often used to produce grape juice processing. Grapes are, usually, crushed fruit or mash passed through a large bore, tubular heat exchanger where it is heated from 50 to 60°C.

This stage, known as the hot break process, is designed to extract a large amount of color and to assist in maximizing the yield. Hot pressing is appropriate for deeply pigmented grapes where maximum color extraction is desired, while the immediate or cold press procedure is necessary to maintain the initial color of light colored grapes.

After extracted, the juice can be concentrated, which is a vital operation of the juice processing industry to reduce transportation costs and storage, as well as to extend the shelf life and provide the supply of juice to the industries during grape off-season.

The most commonly used technology is the vacuum evaporation, in which there is significant reduction of water content of the product [25]. It may be concentrated, as well, by evaporation or freeze concentration, and may be used in grape making and multifruit juices and as a sweetener in other food products and fruit spreads.

Grape juice is concentrated to 55, 65, or 68 °Brix and after the concentration, the product is diluted into single-strength grape juice, multifruit juice, and sparkling juice [26] with a significant reduction of water content of the product [25]. However, processes that use high temperatures can cause loss of thermo sensitive substances, phenolic compounds, color degradation, loss of smell, and can cause cooked flavor to the juice and formation of non-enzymatic browning products such as hydroxymethylfurfural [27, 28, 29]

Due to these changes that might affect the product quality, softer techniques have been proposed for the production of concentrated juice, such as concentration by freezing and membrane separation processes like reverse osmosis.

Among the cultivars used for production of grape juice, *Concord* juice and *Concord* blends have become the standard for quality red grape juice around the world. For white juice, *Niagara* along with *Delaware* and *Catawba* and various *labrusca* blends are gaining in popularity.

Nevertheless, any grape cultivar with acceptable fresh eating quality can be used for juice. In terms of world market, grape concentrate competes with apple and other fruit juice concentrates. It is also used to sweeten jams, jellies, yogurt, frozen fruit desserts, cereals, cookies, and other bakery products.

3.3. Raisins

Raisins are essentially dried grapes. The term has become limited mainly to the dried grapes of a few varieties. Three of which, i.e., *Thompson* seedless, *Black Corinth* and *Muscat of Alexandria*, produce nearly all the raisins of international trade. The method of drying can be natural, golden-bleached, sulfur-bleached and lexia, and the conditions in which the raisins are offered for sale can be layers, loose and seeded, and the size grades (4 crown, 3 crown, 2 crown, and so on). Besides, it can be presented according to different maturity (better and substandard) and quality grades, such as extra standard, substandard, extra fancy, fancy, choice, etc. Mechanical harvesting is commonly employed to process the grapes into raisins. State of California is the biggest producer, contributing 50% to the total world production. Turkey, Australia, Greece, Iran, the Republic of South Africa, and Spain are other important raisin-producing countries [30, 31].

Winkler et al.[1] have given a detailed account of various kinds of raisins, their production methods, handling, and factories affecting their quality: a) size of the raisin berries, b) Hue, uniformity, and brilliance of the color, c) condition of the berry surfaces, d) texture of the skin and pulp, e) moisture content, f) chemical composition, g) presence of decay (rot), mold, yeast and of foreign matter, h) insect infestation.

3.4. Other Products

3.4.1. Grape Spreads

Grape jelly, jam, preserves, butter, or marmalade are made from whole or crushed grapes. The process consists mainly of cooking the grapes and/or their juice in combination with sweeteners and pectins to the proper solids level. There are federal standards that dictate the ingredients, their proportions, and the final concentration of soluble solids. To illustrate this point, the ratios of minimum total soluble solids to fruit sweetener as required for grape jelly, jam, preserves, and fruit butter are 65, 68 and 43 % soluble solids, respectively [26].

3.4.2. Vinegar

The word "vinegar" derives from the Old French word "vinaigre" meaning "sour wine". It is made from red or white wine and is the most commonly used vinegar in Mediterranean countries and Central Europe. As with wine, there are considerable ranges in quality, such as the Balsamic vinegar, a typical product from the area of Modena and Reggio Emilia in the north center of Italy.

According to Italian laws, traditional balsamic vinegar (Aceto Balsamico tradizionale di Reggio Emilia, Italy) (Law No. 93, 1986), is obtained from natural sugar alcoholic fermentation and acetic oxidation of unfermented, cooked, and concentrated crushed grape (must) (≥30 °Brix, according to Italian regulations) of the cultivar *Trebbiano*, *Spergola*, *Lambrusco*, and other local grape-wine varieties. In addition, there are two commercially available balsamic vinegar types on the market: "Aceto Balsamico Tradizionale di Modena" and "Aceto Balsamico di Modena" (industrially made).

The traditional balsamic vinegar must be aged for at least 12 years but it is possible to find some series of casks up to 100 to 200 years old. During the process, microbial metabolism and natural evaporation induce a marked increase in the concentration of derivatives of acetic fermentation, oxidation, and maillard reaction, which impart the typical sensorial characteristics to the final product. Balsamic vinegar is comparable with vinegar made by the traditional "solera" system used for "Jerez" wines in Spain [32, 33, 34, 35, 36].

3.5.3. Canning

Canning is the most commonly used technique to heat and sterilize foods in order to prevent microbiological and enzymatic spoilage. Normally, the seedless kinds, usually *Thompson Seedless*, are used primarily in combination with other fruits for canned fruit salad and cocktail or even alone. Water, fruit juice, sugar, organic acids and carbohydrate sweetener may be added in one of the packing media [37].

4. Nutritional and Functional Aspects

4.1. Chemical Composition of Grapes and Grape Products

As mentioned previously, the chemical composition of grapes varies with species and growing conditions. Nevertheless, the composition of certain nutrients remains constant in the fruits, being modified in their products according to the technological process employed. Thus, the quality of the raw material should be considered since it will be reflected in the originating products.

In general, the fruit is reduced in lipids, proteins (the main amino acids are arginine, proline and glutamine) and dietary fiber and presents higher contents of water, sugars, organics acids and some minerals [38, 39, 40]. Water is the quantitatively most important substance, present around 81 % to 86 % [41], and sugars are the main constituents of the soluble solids, being glucose and fructose the predominant ones. With the advance of ripening, the content of sugar increases while the total acidity reduces [42, 43].

Acidity is due essentially to the tartaric and malic acids, which are predominant, and to the citric and succinic acids, present in minor proportions. Such organic acids render low pH, providing balance between the sweet and acid taste, as well as stimulating the salivary and gastric secretions [44]. Contents of these components may vary according to the grape specie, the conditions of climate and soil of the cultivation area, the ripening stage and to the possible crystallization of the potassium bitartrate [42, 45, 46, 47].

Among the minerals, potassium is highlighted, and as higher it's content, more elevated is the pH of the fruits. Although in lower concentrations, there may be presence of calcium, magnesium, iron, copper, zinc, manganese and lithium [43].

In wine manufacturing, often the must lacks enough sugar content and therefore, has low potential for production of alcoholic wines, which can occur when the grapes do not reach optimum ripeness or when they are soaked with water during periods of rainfall prior to harvest[48]. In such cases, the technique of chaptalization - the addition of sucrose to the must - is commonly used to correct the deficiency of sugar in fruit, but the addition of concentrated grape must has been used successfully in this process [49, 50].

The formation of other substances is another aspect of importance in wine industry. Fermentation process brings about modifications like the production of ethanol, and as consequence, decrease of total sugar content. After water, ethanol is the constituent present in higher amounts in wines, about 10 to 13 % of volume.

4.2. Main Classes of Bioactive Compounds

Phenolic compounds represent constituents of importance in the grape and its products, since they are responsible for the sensory characteristics of color, astringency and bitterness e represent some of the most abundant antioxidants of the human diet [43, 51]. These phytochemicals are synthesized in the plants from the amino acids phenylalanine and tyrosine after environmental injuries like ultraviolet radiation or pathogen aggression, or as mechanism of attraction to pollinator insects, as antioxidants, among other functions [52, 53].

The structure of these compounds consists in diverse hydroxyl linked to aromatic rings, being divided in two distinct groups: the non flavonoids and the flavonoids, where the last consist of a basic structure of two aromatic rings (A and B) linked by three atoms of carbon, forming an oxygenated heterocycle (ring C) [52].

In grapes, the main phenolic compounds are the phenolic acids (derived from cinnamic and benzoic acids) and the stilbenes (resveratrol) – belonging to the class of non flavonoids -, and the flavonoids of the group of flavanols (catechin, epicatechin and epigallocatechin), flavonols (kaempferol, quercetin and myricetin), tannins and anthocyanins [43, 28, 54]. The structure of these compounds is illustrated in Figure 1.

Phenolic acids are the main phenolic compounds of the grape pulp [55]. The peel concentrates the flavonols and in the red and purple varieties, the anthocyanins, reason why the white varieties of grape and its by-products (eg white wine and juice) experiment lower concentrations of phenolic compounds [56, 25, 57]. Flavanols, that constitute the main subunits of the tannins, are mostly present in the seeds and in the bunch stem [58, 59].

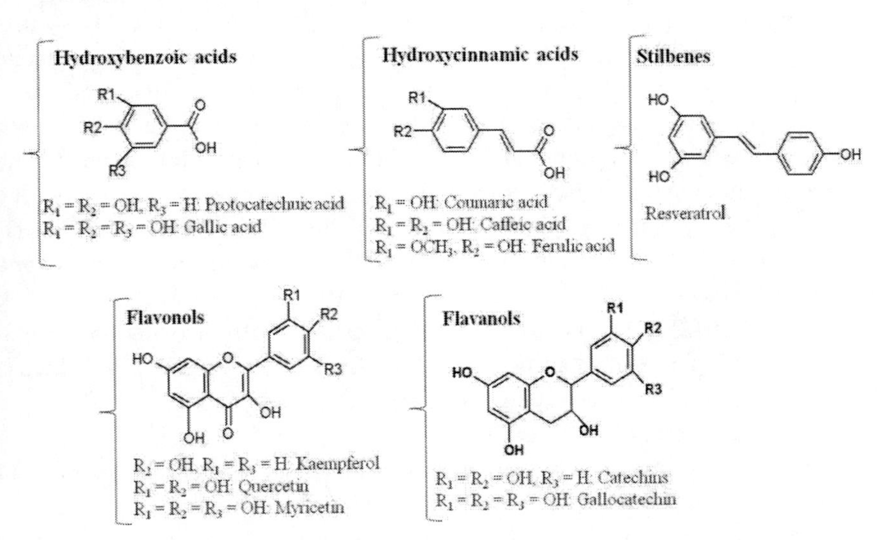

Figure 1. Structure of the main phenolic compounds of red grape varieties and its products.

Resveratrol is a phytoalexin synthesized mostly in the grape peel especially after *Botrytis cinerea* infection [60, 61], and is present in concentrations between 0.19 and 0.90 mg of *trans*-resveratrol.L^{-1} of grape juice, and about 1.3 to 7.0 mg.L^{-1} in wines [62].

Composition of phenolic compounds in grape is dependent on environmental factors, cultivation, as well as the specie used in the product manufacturing. In grape juice, Gollücke et al. [63] observed significant changes in the phenolics composition during different processes. Peonidin and peonidin-3-glucoside comprised the main components before the thermal treatment, and after pasteurization, malvidin and dimethoxy-flavilium were the predominant. The stilbene piceatanol-glucoside became higher after the concentration process.

4.3. Anthocyanins

Anthocyanins comprise the vegetable pigments, water soluble, responsible for the pink, red, violet and blue colors of different flowers, fruits and leafs [64]. In nature, anthocyanins exist in the glucosilated form, where the aglicons are known as anthocyanidins, having the basic structure of cation flavylium [65]. Anthocyanin molecule (Figura 2) is constituted by an aglicone (anthocyanidin), a group of sugars and, frequently, a group of organic acids.

Six types of anthocyanidins are important in foods: malvidin, delphinidin, petunidin, peonidin, cyanidin and pelargonidin, which differ between one another by the number of hydroxyls and to grade of methoxyls present in the ring B[66]. The formation of the glucoside and the higher presence of OCH_3 groups render, generally, reddish color and higher stability to oxidation and heat than the raise of hydroxilations, that therefore provide prevalence of pink and blue colors [25, 63]. Due to the countless patterns of glucosilation, several anthocyanins have been identified. Glucosilations occur mostly with glucose in the positions 3 and 5, being possible with other monosaccharide like galactose, ramnose and arabinose [67].

Grapes represent a rich source of anthocyanins, where the industrial colorant originating from the peels is named “enocyanin”[68]. Of the six anthocyanidins, only pelargonidin is not found in this matrix, although some authors have reported the presence of this aglicone in trace amounts[69]. Regarding the glicosilation pattern, American grapes differ from the Europeans, since the cultivars of *Vitis labrusca* present mono and diglucosides, while *Vitis vinifera*, only 3-monoglucosides [70, 71, 72].

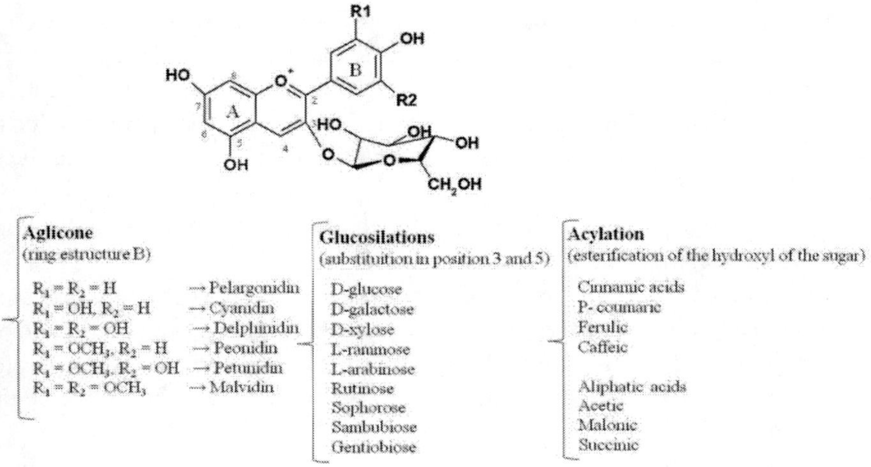

Aglicone
(ring estructure B)

$R_1 = R_2 = H$ → Pelargonidin
$R_1 = OH, R_2 = H$ → Cyanidin
$R_1 = R_2 = OH$ → Delphinidin
$R_1 = OCH_3, R_2 = H$ → Peonidin
$R_1 = OCH_3, R_2 = OH$ → Petunidin
$R_1 = R_2 = OCH_3$ → Malvidin

Glucosilations
(substituition in position 3 and 5)

D-glucose
D-galactose
D-xylose
L-ramnose
L-arabinose
Rutinose
Sophorose
Sambubiose
Gentiobiose

Acylation
(esterification of the hydroxyl of the sugar)

Cinnamic acids
P- coumaric
Ferulic
Caffeic

Aliphatic acids
Acetic
Malonic
Succinic

Figure 2. Chemical structure of the anthocyanins.

Many studies have identified glycosides of delphinidin and cyanidin as the predominant anthocyanins in cultivars of *V. labrusca*, however, it was proposed that malvidin-3,5-diglucoside were used for quantification of anthocyanins in stored products, as it is considered more stable [73].

The typical red/violaceous color is associated to the higher quality of grape juice [74] and the composition of anthocyanins in this beverage is affected during the processing steps like pressing, heating, enzymatic treatments, pasteurization and storage [2]8. As well, anthocyanins in wine undergoes several modifications depending on pressing, fermentation, clarification, stabilization, bottling, aging among others, that affect beneficially or not the stability of these structures [25].

A phenomenon that might occur is the decrease in the concentration of monomeric anthocyanins, due to the polymerization reaction, generating polymeric pigments that are more stable to the changes of the medium [75, 28]. Another mechanism of protection is the copigmentation, where the anthocyanins bound, probably by hydrogen bonds, with tannins, proteins, non-anthocyanin flavonoids, organic acids, alkaloids, polysaccharides and metal ions [76], where the cation flavylium as well quinoidal base may participate in this lost preventing color process.

The antioxidant activity of the anthocyanin *in vivo* is related to the deficiency of electrons in the nucleus flavylium and to the presence of free hydroxyl, resulting in high reactivity. However, this condition may bring about reactions in the food matrix that generally involve the pigment loss of color, which is undesirable in the processing of fruits and vegetables [77, 78].

Besides cation flavylium, other structural forms (Figure 3) may occur in aqueous solution depending on the medium conditions, like pH variations, temperature, light, presence of oxygen, metal ions, ascorbic acid, additives among others [65], and these interferences can affect the stability of anthocyanins and cause its degradation, which directly influences the resulting color of these pigments.

The form of the cation flavylium is prevalent in low pH (less than 2) and gives reddish color. With the pH increased to about 3, the flavylium cations begin the loss of protons and the quinoidal base is formed, with purple coloration, as well as the ionized quinoidal base, which determines the blue or violet color.

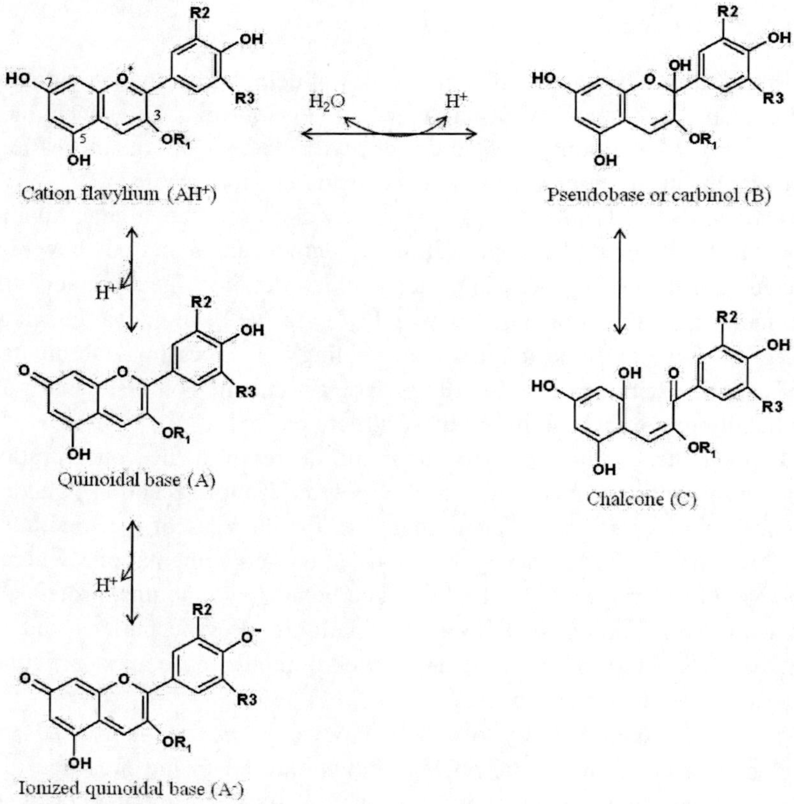

Cation flavylium (AH^+)

Pseudobase or carbinol (B)

Quinoidal base (A)

Chalcone (C)

Ionized quinoidal base (A^-)

Figure 3. Structural forms of anthocyanins in aqueous solution.

Alongside, the cation flavylium hydration occurs with the formation of pseudobase or carbinol, colorless, and chalcones, pale yellow. If the time of exposure to high pH is extensive, there is a high proportion of chalcones and loss of color intensity [79]. Thus, for general application in the food, pH between 1.0 and 3.5 gives more stability to anthocyanins. The average pH found in grape juices and wines is about 2.8 to 3.8, making these products naturally prone to increased anthocyanin stability [77].

Another important factor for the destabilization of anthocyanins is the light, which effect is strongest when combined with the presence of oxygen [64]. Heat can cause degradation of anthocyanins, possibly by breaking the glycosidic bond and formation of chalcone, so in the processes that use high temperatures are recommended short times of exposure to avoid the loss of pigments, situation that occurs in the regular processes of grape juice extraction [65].

Sulfites or metabisulfites used in foods as antioxidants and antimicrobials, and in wine production, also interfere with the form of anthocyanins, which can cause irreversible bleaching when used at high concentrations, above 10 g.kg- [1 67].

4.4. Grapes, Grapes Products and Human Health

Benefits to human health provided by red grapes achieved a highlight with the French Paradox theory, where a low mortality rate was observed in the French population that consumed a diet rich in saturated fat but also presented an elevated consumption of red wine [80]. Several researches were conducted in order to evaluate the effects of red wines in human health, and part of the protective effects of red wine were attributed to the alcohol capacity of raising the high density lipoprotein (HDL), inhibit platelet adhesion and enhance absorption of the phenolic compounds [81, 82]. Nevertheless, some studies indicated that the benefic effects of wine were achieved independently of the alcohol, and were attributed to the phenolic compounds of this beverage [83].

Thus, recommendations for the consumption of an alcoholic beverage like wine as a cardio protective strategy is contradictory and must be evaluated individually, since ethanol is responsible for some injuries to health, like decreased absorption of some vitamins, which can lead to Wernicke-Korsakoff syndrome; liver and stomach diseases; alcoholic dependence; pancreatitis, fetal alcohol syndrome etc. [84, 85].

Table 1. Experimental studies with grape products in humans and animals

	Authors	Experimental design	Results	Conclusions
Experiments with human	*1*	12 ♂ and 3 ♀with coronary artery disease and average age of 62.5 years consumed nearly 7.7mL.kg^{-1}.d^{-1}of red Concord grape juice for 14 days	Improvement of VMF compared to baseline; decreased susceptibility of LDL oxidation; improvement of endothelial function even in patients using statins and antioxidant vitamins. There was a slight increase in cholesterol and triglycerides	Ingestion of grape juice can prevent cardiovascular risk by mechanisms independent of alcohol, such as increased production of nitric oxide
	2	5 ♂ and 5 ♀, healthy , aged between 26 and 58 years consumed 5 – 7,5mL.Kg^{-1} of grape juice (Concord), orange juice or grapefruit juice, one at a time, for 1 week. At the end of each intake of juice, it followed a week without eating any of the juices.	77% reduction in platelet aggregation compared to baseline with consumption of grape juice. Platelet aggregation in response to increased concentration of collagen was inhibited by 21% with grape juice, which was not observed with orange juice or grapefruit	Grape juice inhibited platelet activity, which did not occur with other juices. This was attributed to the almost 3 fold higher concentration of polyphenols in grape juice, as well as classes of phenolic compounds present
	3	Incubation of platelets with diluted grape juice led to inhibition of platelet aggregation, increased platelet-derived nitric oxide and reduction of superoxide anion in vitro. To confirm these experiments, 20 healthy individuals (12 ♂ and 8♀) aged between 20 and 45, consumed 7 mL.kg^{-1}.d^{-1} of Concord grape juice for 14 days	Supplementation of grape juice inhibited platelet aggregation and superoxide anion production and significantly increased production of platelet-derived nitric oxide	Experiments in vitro and ex vivo with grape juice demonstrated the beneficial effects of this drink, as the suppression of platelet-mediated thrombosis, which occurred independently of alcohol consumption
	4	32 healthy adults aged between 27.5 and 28.4 years consumed for 2 weeks 400 IU per day of α-tocopherol (6 ♂ and 11♀) or 10 mL.Kg^{-1}.d^{-1} of Concord grape juice (7♂and 8 ♀)	Supplementation with grape juice increased the antioxidant capacity of serum and the resistance of LDL to oxidation, which were comparable to the dose of α-tocopherol. Only the juice reduced the protein oxidation, and was observed a small increase in triglycerides	Grape juice ingested at recommended doses was able to increase the antioxidant capacity and reduce protein oxidation and of LDL.

	5	16 adults (8 ♂ and 8 ♀) with a mean age of 51.6 years and increased LDL received 500 $ml.d^{-1}$ of red grape juice or 250 $ml.d^{-1}$ of Pinot Noir wine for 14 days followed by a rest of equal duration and introduction of the new treatment	FVM raised with grape juice and red wine. Only the wine increased endothelium-dependent vasodilation. Only the juice reduced ICAM-1. There were no changes in plasma lipids, glucose or platelet aggregation with wine or juice	The effects of juice and wine in the vasodilation appear to be due to phenolic compounds, which can decrease the production of endothelin and in the case of juice, suppress activation of NfκB , reducing ICAM-1
	6	26 patients (13 ♂ and 13 ♀, average age 62 years) on hemodialysis and 15 healthy subjects consumed 100 $mL.d^{-1}$ of concentrated grape juice (68°Brix) of Bobal grape (*V.vinifera*) for 14 days. 12 additional patients undergoing hemodialysis were considered control	Consumption of grape juice reduced LDL, apo B and oxidized LDL and increased HDL, apo AI and TEAC in hemodialysis patients and healthy, but not in the control group. Plasma concentrations of glucose, uric acid, albumin and protein did not change with the supplementation of juice	Supplementation of grape juice improved the plasma lipoprotein profile and reduced the concentrations of MCP-1, which is important in hemodialysis patients who are at high risk of developing cardiovascular disease
	7	18553 households were assessed to determine whether there was an association between mothers' light drinking during pregnancy and risk of behavioral problems, and cognitive deficits in their children at age 3 years. Two sweeps occurred: when the cohort members were aged 9 months and 3 years. Behavioral problems were indicated by scores falling above defined clinically relevant cut-offs	Children born to light drinkers (1–2 drinks per week or per occasion) were less likely to score above the cut-offs, and the ones born to heavy drinkers (7 or more units per week or 6 or more units per occasion) were more prompted to score above, compared with children of abstinent mothers. Boys born to light drinkers were less likely to have conduct problems, hyperactivity and presented higher cognitive ability, and the girls , less likely to have emotional symptoms and peer problems compared with those born to abstainers.	Children of light drinkers were not at increased risk of clinically relevant behavioral difficulties or cognitive deficits compared with children of abstinent mothers. On the opposite, heavy drinking during pregnancy appears to be associated with such problems.

Table 1. (continued)

Authors		Experimental design	Results	Conclusions
	8	149 773 subjects were divided into groups according to alcohol consumption: never, low (<10 g/day), moderate (10–30 g/day) , high (>30 g/day), former drinkers.	Moderate male drinkers displayed conditions associated with lower cardiovascular risk, including low body mass index, heart rate, pulse pressure, fasting triglycerides, fasting glucose, stress and depression scores together with superior subjective health status, respiratory function, social status and physical activity. As well, moderate female drinkers presented low waist circumference, blood pressure and fasting triglycerides and low-density lipoprotein-cholesterol. Importantly, few factors were causally related to alcohol intake.	Moderate alcohol drinkers showed a more favorable clinical and biological profile, when compared with nondrinkers and heavy drinkers. Thus, moderate alcohol consumption may represent a marker of superior health status and lower risk of cardiovascular disease.
Experiments with animals	*9*	Golden Syrian hamsters (9 in each group) received a diet rich in cholesterol and saturated fat to induce formation of foam cells. The control groups received either water or 6.75% ethanol, and the other groups red wine, red wine without alcohol or diluted Concord grape juice	Ethanol, wine, alcohol-free wine and grape juice reduced atherosclerosis. Wine and juice had a hypocholesterolemic effect compared to control, except alcohol. Between the beverages rich in polyphenols, only the group of grape juice showed increased HDL	When compared to the alcohol-free wine and standardized to the dose of polyphenols, the benefic effects of red wine were attributed entirely to the polyphenols. It was estimated that the grape juice was more effective than wine or alcohol-free wine in inhibit atherosclerosis and improve the lipids profile and antioxidants.
	10	45 rats (Fischer 344) with 19 months of age were divided in 3 groups that received 0%, 10% or 50% of grape juice.	The rats that consumed 10% and 50% of grape juice achieved better performance in the behavioral tests. The 10% group presented increase of the cognitive performance and the 50% group improved the performance in motor tests (balance, coordination and strength)	Results suggests that the phenolic compounds of grape juice may be benefic in revert the brain aging and reduce the risk of neurological diseases resulted from oxidative stress and inflammation

	11	20 rabbits (10 in control group and 10 in the treated) received hypercholesterolemic diet for 48 days and after, the treated group consumed 225 $mL.d^{-1}$ of Concord grape juice for 48 days, but continued receiving the diet rich in cholesterol	In the 96º day the treated group showed significant reduction of the platelet adhesion, the development of atheroma, of the total cholesterol and arterial pressure, which was not observed in the control group	Authors reported that the polyphenols of grape juice have impact in many physiological processes, but careful must be taken in the recommendations for human health
	12	The antibacterial properties of resveratrol towards 17 *H. pylori* strains was tested, as well as the inhibition of *H. pylori* urease by resveratrol and red wine	An inhibition of all the strains by resveratrol was observed in concentrations of 25 to 100 μg/mL. In addition, resveratrol and red wines presented an inhibitory effect on *H. pylori* urease activity	Resveratrol and red wine may have therapeutic potential for application in illnesses caused by *H. pylori.*

FVM= Flow-mediated vasodilation.

[1]Stein et al. (1999) [94, 2] Keevil et al. (2000) [95, 3]Freedman et al. (2001) [92, 4] O'Byrne et al. (2002) [89, 5]Coimbra et al. (2005) [85, 6] Castilla et al. (2006) [91], [7]Kelly et al. (2009) [87, 8] Breuil et al. (2010) [60, 9] Vinson, Teufel e Wu (2001) [83, 10] Shukitt-Hale et al. (2006) [100, 11] Shanmuganayagam et al. (2007) [102, 12] Paulo et al. (2011) [101]

Most of these effects, actually, are mainly observed in heavy drinkers (over 30 g of alcohol per day). Low (<10 g/day) to moderate alcohol consumption (10-30 g/day) by healthy adults has shown beneficial effects, compared to abstinence and heavy drinking [86]. Other data corroborates the light ingestion of alcoholic beverages in the group of pregnant women, where ethanol in the diet is of great concern, because of the problems that can be caused to the fetus [87].

Red grape juice represents a nonalcoholic alternative to wine and can be consumed even by people with diseases like hepatitis, children and adolescents, elderly, pregnant women and abstainers [28]. The juice consumption should be careful in diabetic subjects, for the high glycemic index and glycemic load, or by people with hyperuricemia or athletes, since fructose can increase the degradation of adenine nucleotide, and glucose stimulates the secretion of insulin, hormone that can lead to reabsorption of uric acid in the kidneys, increasing serum concentrations of this substance [88].

Since the frequent intake of foods rich in phenolic compounds is associated with decreased incidence of chronic diseases [82, 89], a series of experimental studies confirming the effects of ingestion of grape juice and red wine (included in alcoholic beverages) can be seen in Table 1.

For the polyphenols exert their antioxidant functions *in vivo* they need to be properly absorbed and metabolized. O'Byrne and colleagues [89] documented the increase in concentration of these compounds in the free and conjugated form in plasma of subjects after consumption of grape juice, which were maintained in the bloodstream for up to 12 hours. Bitsch et al. [90] observed higher absorption of anthocyanins in grape juice compared to red wine, probably due to the synergistic effect of glucose and was reported by Galvano et al. [78] that anthocyanins can be absorbed intact, rather than just their aglycones. Castilla et al. [91] observed elevated plasma quercetin until three hours after ingestion of red grape juice. Other studies showed increased plasma antioxidant capacity after consumption of this drink [92, 93].

Properties of grape products, especially the red varieties, in decrease the risk of diseases and the therapeutic effects on cardiovascular diseases, neurological disorders and even cancer[54] are attributed to the phenolic compounds, which are capable of acting in hydro and lipophilic environments [89] and exert beneficial effects through different mechanisms of action such as antioxidant activity on low density lipoprotein (LDL) and against superoxide anion, possibly as hydrogen donors [94, 95, 89, 78, 96]; inhibition of nuclear factor κB activation [85]; decrease the expression of pro-inflammatory enzymes like cyclooxygenase-2 and iNOS [97]; chelation of metal ions [78]; reduction of inflammation markers of and monocyte adhesion such as ICAM-1 and MCP-1 [91]; inhibition of platelet aggregation [92, 61]; induction of nitric oxide production and decreased production of endothelin, promoting vasodilation [92, 85, 98, 99]; improved function of the endothelium [94] and brain [100].

5. By-Products of Grape Processing

The constant growth of world population, which forecast to 2050 is nine billion people [103], challenges the food supply and compel industries, governments and academic community to seek alternatives to an increased and better use of natural sources, as well as giving special attention to waste due to the variety of possible uses. Estimates point out that almost half of all food is lost before and after reaching consumers, including harvest, post-harvest, transportation, processing, storage, packing, preparation etc [104]. Nevertheless, nowadays environmental impacts arising from production activities in all sectors of society are widely discussed and brings about more specific legislation of environmental responsibility by the governments.

Concerns about food losses stimulated FAO (Food and Agriculture Organization of the United Nations), in 1945, to establish reductions on this issue, and in 1974, the first World Food Conference marked the identification of a problem to solve hunger: reduction of post-harvest losses. Following this idea some actions were conducted within these years, notwithstanding, the poor results obtained indicated the strategies mistakes and a more holistic approach, suggested by Groullead [105], and was adopted in order to solve the problem or at least minimize it. Accordingly, concepts of supply chain management with no-residues generation applied holistic view of the process, evaluating all impacts produced through the whole chain and redirecting and resizing processes with a view to minimize the wastes until reaching an integrated productive process guided by environmental protection since the initial steps of product development, with high quality value, safety and efficiency [106].

Following hierarchy steps to reaching "green production" there are processes of short, medium and long time, being in order of preference to avoid, reduce, reuse, recover, treat and dispose the wastes. Food industries, especially of fruits and vegetables, generate large amounts of by-products and residues, which present high biochemical and chemical oxygen demand, increasing costs for the final dispose [106]. Still, those may be valuable for its high nutritional and functional content, emerging as an interesting by-product.

Fruits and vegetables waste became a major issue, due to the increased consumption of these products, especially industrialized forms such as juices and derivatives, in view of the epidemiological studies demonstrating the benefits to health of a diet rich in fruits and vegetables.

Regarding grape, is the second most produced fruit in the world and almost 80% of the global harvest is intended for the processing of wine, juice and derivatives [107], generating large amounts of solid and liquid residues (Figure 4). Among them, grape pomace or marc, grape stalkand wine lee are the main solid by-products and waste generated in wine processing, although exist winery-sludge, vinasse and winery wastewater. Estimates show that for every 100 liters of white vine, 31Kg of by-products are generated whilst for red vine this value is of 25Kg (Costa and Belchior,1972 apud [108]).

Grape pomace is composed by peel and seed, yet depending on the industry may be totally or partially de-stemming, being the largest by-products in quantity, around 20% of the whole weight of processed grape [109]. Phenolic compounds, especially flavonoids and stilbenes, are the main functional compounds present in the skins and seeds and their relevance is associated with biological activities arising mainly from the antioxidant capacity.

Grape stalk is the skeleton of the grape bunch and is basically constituted by fiber, lignins and celluloses, and minerals nutrients as potassium and nitrogen [111]. Lignins are phenolic compounds belonging to the non-flavonoid macro class, presenting biological properties too.

Lees or winery –lee is the residue disposed on the bottom of the container after fermentation of wine, during storage or after authorized treatments (ECC regulation n° 337/79 *apud* Perez-Serradilla and Luque-de-Castro [112]), and are also considered as lees the residues deposited on grape must recipients. It is characterized by high content of wine (70 a 90%), tartaric substances, yeast and other compounds [108]. Although this is not yet a common practice in the world, European Union, that comprehends important wine producer's countries, established in 1999 an agreement that requires wineries allocate grape pomace and wine-lee for distilleries, decreasing the quantity of them but generating vinasse and grape marc exhausted [113]. However, this policy hardly reaches smaller wine producers neither others countries, being they donate to use as animal feed or fertilizers, the routine adopted to avoid accumulation of waste and lowing costs to disposal.

Albeit the specific characteristics of processes will be influence the quantity of wastewater generates, Airoldi et al. [114] estimate that it corresponds at 75 % of whole residue produced in wineries with a 130 to 150 kg for each hundred liters of wine. Winery-sludge is a solid residue formed in the walls of recipients and recovering with hot or vapor during plant cleaning, and its actual valorization is associated at tartaric substances.

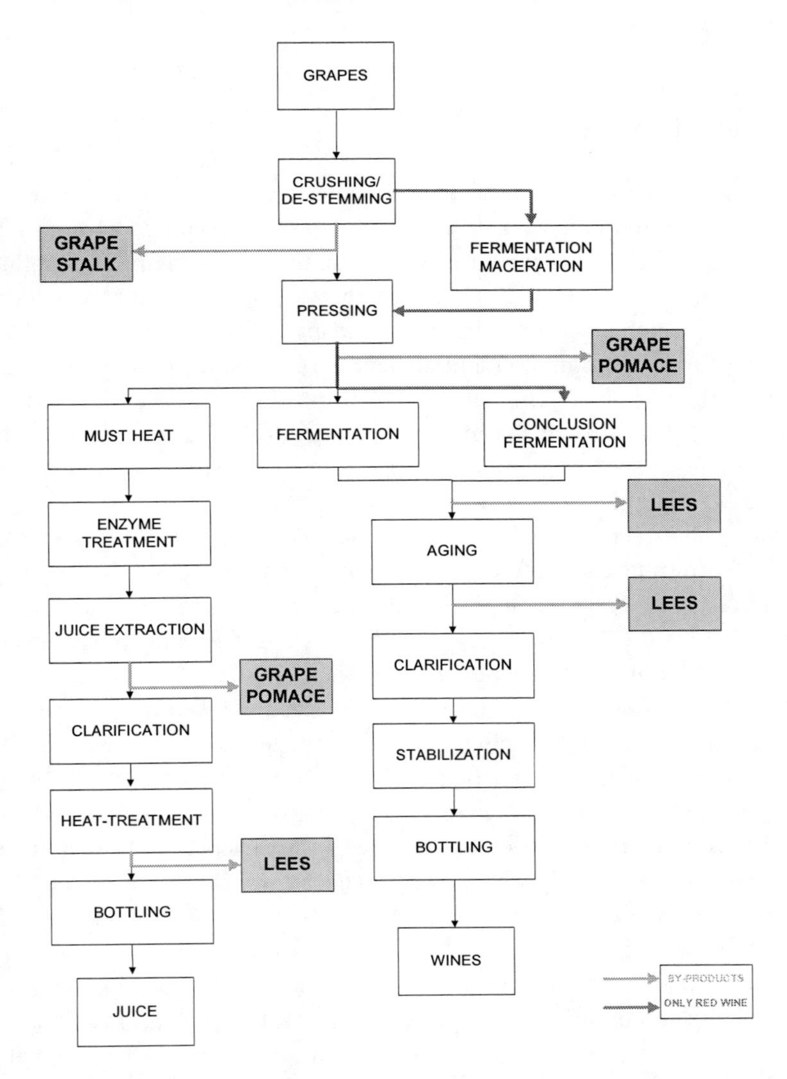

Figure 4. Simplified scheme of grape processing and generated by-products (adapted from [110, 41])

Differences in composition residues as regard from processing conditions, type of by-product, origin of by-product (juice, white or red wine) and cultivars grape used will influence on the direction of final destination current and perspective of them. Economical viability of these actions is becoming reality, specially for products with health benefits, being possible find in the USA market products base of grape seed, grape skin, anthocyanins extracts and others extracts [115].

5.1. Grape Pomace

The grape marc become a environmental issue to be solve by industry, because beside the pretty quantity generates in a short period of time - 16 to 20% of grape weight processed in three months crop - it presents pollutants characteristics as low pH, high phenolics content, antibacterial compounds and phytotoxins, which resist a biological degradation [113].

One of the most common and ancient destination of grape pomace are to feed animals and plant. The use as fertilizers is not completely established; Ferer et al. [116] verified the efficient of organic fertilizer from fresh grape waste composting over corn growing when compared with commercial fertilizer. Nevertheless Bustamante et al. [117] identified a negative correlation between high initial phenolics content with the decrease of mineralised nitrogen and increase of nitrogen immobilization, however grape wastes presented a high value in organic matter and important macronutrients for growing plant [113]. And so, the treatments to reduce phenolic compounds seems be a possible to cover come this problem.

Even de-seeded grape pomace was observed a low metabolized energy and lower availability of protein that found associated with acidic dietary fiber, limiting the use of this waste for feed ruminant's purpose [118]. Though the including grape pomace in diet of cattle is regarding the positive effects on animal health due to phenolics contents [119] or enriched milk with bioactive compounds, once secondary metabolites can be transferred to milk [120], in this sense the supplementation of phenolics extracts without proanthocyanidins would be an option.

Despite this, a composition rich in fibers and bioatives compounds could be interesting and promising sources of functional compounds, raw-material to new products or another nobler destination. These characteristics vary with the processing that led to grape pomace (fermented and no-fermented), cultivars employed, de-stemming step, among others. Between the grape pomace generates, that obtained from white wine production is the less exhausted one, because it don´t suffer temperature, enzymatic and alcoholics actions, while the red wine grape marc is the most poor in bioactive compounds of them.

Constituted by seeds, peels and stem grape marc presents an interesting composition on flavonoids – anthocyanins, flavanols and flavonol – and non-flavonoids such phenolics acids and stilbenes. Phenolics compounds are one of the most numerous and widely spread secondary metabolites of vegetables, derived of phenylalanine and tyrosine amino acids shynthesized normally by plants and in stress response[53]. They play different roles in vegetable as phytoalexins, contributing for pigmentation, attractive to pollinators, as antioxidant and protection of wounding, UV radiation, and infections among others [97]. In the food and beverages they can response for color, astringency, bitterness, odor, flavor and oxidative stability [55].

For long time associated as antinutritional actions phenolics compounds have aroused interest because positive association of them with chronic diseases (neurodegenerative, cancer and cardiovascular prevention that the oxidative stress is a primary causes [121, 122]. Plural pharmacology phenolics effects as bactericid, antiviral, anti-inflamatory, anti-carcinogenic, hepatoprotective, antithrombotic, vasodilatory actions which depends on antioxidant capacity, free radical scavenging, chelation of metallic ions with redox activity, genic expression modulation and signalizing cells interaction [97, 123].

Considering these functional compounds in pomace, grape seeds presents higher content of phenolics – 5 to 8 % - than skins [115] - 1 to 2%- , and a rich non-colored polyphenol extracts maybe a promissory product due widespread applicability. Corroborated these data, Montealegre et al. [124] evaluated seventy grape pomace samples comprehending six white and four red grape varieties, and founding flavonoids (flavonols and flavan-3-ols) contents in fresh grape varying 79 to 180 mg/Kg for white and 222 to 253 mg/Kg for red skins while in seeds the flavan-3-ols contents achieving 1390 mg/Kg for white and 870 mg/Kg in red grape seeds. Rockenbach et al. [125] to evaluate Vitis vinifera and Vitis labrusca pomaces observed the same profile, higher values of total flavanols in seeds (6.812mg/100g dry weight) than skins (252 mg/100g dw) that represents a seed extracts twice stronger in antioxidant capacity (8.281 μmol Trolox equivalent/100g dw) than skins one (3.640 μmol Trolox equivalent/100g dw). And tin both studies, proanthocyanidins are the most abundant flavanol.

Notwithstanding grape seed contain only 10 to 20 % of lipids, its oil has a high nutritional quality due to the fatty acid profile, and when extract as virgin oil some biological properties can be observed as consequence of tocoferols and phenolic contents. Another interesting characteristic is high smoke point [126], even only 10 % is saturated fatty acids and 90 % unsaturated, 47 to 60 % of linolenic acid and 9 to 17 % of oleic acid[127].

Red grape skins have been evaluated as a source of natural colorants, that as consequence of consumer demand by natural feedstuff, improvement and innovative their use by adding the multifunctional additives concepts. Classes of natural colorants such as anthocyanins, carotenoids and betalains that posses antioxidant capacity[128] can be classified as multifunctional additives [129]. Application of the extracts is the determinant factor of extraction method choice, for food the main processes comprehended alcoholic, enzymatic and aqueous-sulfite extraction. As mentioned above, the origin of grape pomace influence over composition of bioactive compounds and in consequence over the extracts of them, Gomez-Plaza et al. [130] achieve highest yield extraction of anthocyanins in rosé pomace (1,4 g/Kg pomace) than red wine pomace (0,5 g/Kg pomace).

Others factors such as time, temperature, solvent concentration, solid:liquid rations, pH and the target molecules affected molecules extraction, and can be considering to define the more appropriate condition to obtain the extract. Lapornik et al. [131] observed that for anthocyanins time of extraction influence negatively after twelve hours and ethanol 70 % was the best solvent, while the extraction of total phenolics was favored by the time and use of methanol 70 % as solvent.

Novel technologies as ultrasonic, high pressure hydrostatic and pulse electric fields when applied combined at ethanol solution at grape marc extraction promoting increase in yield of bioactive compounds (at least twice) and in the specific case of anthocyanins a selective recovery was observed, high pressure hydrostatic was better for acylated anthocyanins while monoglucosilated molecules for pulse electric fields [132].

Antioxidant dietary fiber is defined by Saura-Calixto et al. [133] as a supplement with more than 50% of dietary fiber and at least antioxidant capacity equal a 50mg vitamin E. In this sense, Llobera and Cañellas [134] evaluating pomace and stem finding 74 % and 77 % dietary fiber level at pomace and stem, respectively, and high antioxidant capacity suggesting these wastes as a good source to antioxidant dietary fiber, that can be used to prevent lipid oxidation in minced fish frozen [135].

Besides recovery compounds, grape pomace have been tested as raw material in enzymes production by solid state fermentation [136, 137], Monascus pigments production by fermentation in a medium contained grape pomace powdered [138], for biogas production [139], bio-oil production by pyrolysis of grape pomace and grape skins [140].

5.2. Grape Stalks

As mentioned previously, stalks are the skeleton of grape of the grape and corresponding as 3 to 6 % of grape processed weight basically a lignin-cellulosic biomass, being 34 % of lignin, 36 % cellulose, 24 % hemicelluloses and 6 % of tannins [141], that can be a good carbon source been evaluated for biogas production [142].

On the other hand, this waste was considered a source of dietary fiber, due to high content of polysaccharides. And besides the use as antioxidant dietary fiber [134], stalks were evaluated for nutritional benefits once showed good performance to swelling, fat adsorption and water retention [143].

Following the world worry about environmental and pollution problems, lignin cellulosic components of grape stalks showed ability to sorbet chromium (Cr III and Cr VI) in acid solution and even presented the beneficial property to convert chromium hexavalent - most toxic- in trivalent one [144]. Different wastes have been tested as source of activated carbon, universal adsorbent can be applied in different medium, and Deiane et al. [145] submitting grape stalks at two different processes concluded a chemical the most appropriate to obtained activated carbon because the high minerals content difficult carbon fixation.

Composting process allow organic matter to be reincorporate at the land in the better manner than only disposal food processing by-products. As grape stalk comprising a large amount of lignin cellulosic and considerable minerals nutrients rate (nitrogen and potassium) it use as fertilizer have been considered. However some works suggesting a composting as a pre-treatment for used it, due to its chemical characteristics [146, 111, 113].

Recovering bioactive components is another way for added value to wastes, and in the case of stalks the interest is focusing in phenolics compounds, as for many health human effects as for antioxidant capacity that allow to be applied in different systems, such as food, cosmetic, pharmaceutics forms as a protector to oxidation.

Many natural sources of antioxidant have been evaluated, with special attention for by-products in the last time, but vary extraction processes have been tested [147, 148] and kinetics modeling aspects are considered [149] to achieve yield, purity, safety, and antioxidant at considerable level.

5.3. Lees

Lees are formed during the processes and are characterized for the residue deposited at the bottom of recipients that containing wine or grape must, and represents 5 to 8 % wine volume depending on the production techniques [108].

As presenting a high level of wine (70 to 90 %) and tartaric substances (2,5 to 4 %) this residue is sent to alcohol distilleries for recovery of alcohol and calcium tartrate, and the generated vinasse still remains interesting substances (nutrients, phenolics and tartarics) [150].

Beside tartrate, phenolics compounds can be extracted from wine lees. Pérez-Serradilla and Luque-de-Castro [112] verified that microwave-assisted extraction achieved the same yield in a very short period of time compared to conventional extraction of dried and milled wine lee centrifugation solid residue. Yeast present in wine lees has low nutritive value for use in animal feed, but can be composting and applied for agricultural purposes [113].

Conclusion

The constant increasing in production of grapes and derived products is verified mainly by the cultivation of vines in different parts of the world, in which the number of different cultivars and variability of climatic conditions raises the diversity of commercial products.

Besides, the strong investments in research, technology of production (such as the use of membrane separation processes in the production of wine and juices), and the trend for consumption of foods that present functional properties and sensory quality have also influenced positively the enhancement of production, processing and ingestion of grape-derived products.

This chapter also meant to show that the high competitiveness of this productive chain predicts the environmental sustainability, which is being chased especially by the use of residues that can also represent an alternative in the quest for greater economic output and production of processed raw material.

REFERENCES

[1] Winkler, AJ; Cook, JA; Kliewer, WM; Lider, LA. *In General Viticulture*, 2nd Ed. Berkeley: University of California Press 1974.

[2] This, P; Lacombe, T; Thomas, MR. (2006). Historical origins and genetic diversity of wine grapes. *Trends in Genetics,* 22, 511-519.

[3] McGovern, PE. *Ancient wine: the search for the origins of viniculture.* Princeton: University Press. 2004.

[4] Fehér, J; Lengyel, G; Lugasi, A. (2007). The cultural history of wine - theoretical background to wine therapy. *Central European Journal of Medicine,* 2, 379-391.

[5] Zohary, D, Hopf, M., *Domestication of Plants in the Old World.* Oxford: Clarendon Press, 1993.

[6] Manen, JF; Bouby, L; Dalnoki, O; Marinval, P; Turgay, M; Schlumbaum, A. (2003). Microsatellites from archaeological *Vitis vinifera* seeds allow a tentative assignment of the geographical origin of ancient cultivars. *Journal of Archaeological Science,* 30, 721-729.

[7] Margaritis, E. & Jones, M. (2006). Beyond cereals: crop processing and *Vitis vinifera L.*: ethnography, experiment and charred grape remains from Hellenistic Greece. *Journal of Archaeological Science,* 33, 784-805.

[8] Sadori, L., Susanna, F., Persiani, C. (2006). Archaeobotanical data and crop storage evidence from an early Bronze Age 2 burnt house at Arslantepe, Malatya, Turkey. *Vegetation History and Archaeobotany,* 15, 205-215.

[9] Miller, N.F. (2008). Sweeter than wine? The use of the grape in early western Asia. *Antiquity*, 82, 937-946.

[10] Figueiral, I; Bouby, L; Buffat, L; Petitot, H; Terral, J-F. (2010).Archaeobotany, vine growing and wine producing in Roman Southern France: the site of Gasquinoy (Béziers, Hérault). *Journal of Archaeological Science*, 37, 139-149.

[11] Zohary, D. Domestication of the Grapevine *Vitis vinifera L.* in the Near East. In: Mc Govern, P.E. et al. The origins and Ancient History of Wine. Amsterdam: Gordon and Breach, 1995.

[12] Terral, J-F; Tabard, E; Bouby, L.; Ivorra, S; Pastor, T; Figueiral, I; Picq, S; Chevance, J-B; Jung, C; Fabre, L; Tardy, C; Compan, M; Bacilieri, R; Lacombe, T; This, P. (2010). Evolution and history of grapevine (*Vitis vinifera*) under domestication: new morphometric perspectives to

understand seed domestication syndrome and reveal origins of ancient European cultivars. *Annals of Botany,* 105, 443-455.

[13] Reish, BI; Pratt. Grapes. In: Janik, J; Moore, JN. *Fruit Breeding, vol. II, Vine and Small Fruits,* Nova York: Widely, 1996.

[14] Arnold, C; Gillet, F; Gobat, M. (1998). Situation de la vigne sauvage *Vitis vinífera* subsp. *silvestris* en Europe. *Vitis,* 37, 159-170.

[15] Arnold, C; Schnitzler, A; Douard, A; Peter, R; Gillet, F. (2005). Is there a future for wild grapevine (*Vitis vinifera* subsp. *silvestris*) in the Rhine Valley? *Biodiversity and Conservation,* 14, 1507-1523.

[16] Vine, RP; Harkness, EM; Linton, SJ. *Winemaking from grape growing to marketplace*, 2 ed. New York: Springer Science+Business Media Inc. 2002.

[17] Kanellis, AK.; Roubelakis-Angelakis, KA; Grape. In: Seymour, GB; Taylor, JE; Tucker, GA, *Biochemistry of fruit ripening*. London: Chapman & Hall, 1993.

[18] Boulton, RB; Singleton, VL; Bisson, LF; Kunkee, RE. *Principles and Practices of Winemaking*. New York: International Thomson Publishing, Chapman & Hall., 1996.

[19] Peynaud, E; Ribéreáu-Gayon, GP. The grape. In: Humlme, AC. *The Biochemistry of Fruits and Their Products*, vol. 2. London: Academic Press,.1971.

[20] Morris, JR; Grape juice: influences of preharvest, harvest, and postharvest practices on quality, In: Harold, EP. *Evaluation of Quality of Fruits and Vegetables*, Westport: AVI Publishing, 1985.

[21] Amerine, MA; Berg, HW; Cruess, WV. *The technology of wine making*. 3 ed. Westport: The AVI Publishing Company Inc., 1972.

[22] Bucelli, P; Costantini, EAC. Wine Grape and Vine Zoning (Chapter 26). In: Costantini, EAC. *Manual of Methods for Soil and Land Evaluation*. Enfield: Science Publishers, 2009.

[23] Jackson, RS. *Wine Science: Principles, Practice, Perception*. 2 ed. California: Academic Press, 2000.

[24] FAO. MAJOR FOOD AND AGRICULTURAL COMMODITIES AND PRODUCERS [online]. 2011 [2011/ 07/30]. Available from: *http://www.fao.org/es/ess/top/commodity.html?lang=en&item=560&year=2005*

[25] Belitz, HD; Grosch, W; Schieberle, P. *Food Chemistry*. 3. ed. Springer: Garching, 2004.

[26] Morris, JR; Striegler, RK. Grape Juice: Factors That Influence Quality, Processing Technology, and Economics. In: Barrett, DM; Somogyi, LP;

Ramaswamy, HS. *Processing Fruits: Science and Technology,* 2 ed. Nova York: CRC Press, 2005.

[27] Petrotos, KB; Lazarides, HN. (2001). Osmotic concentration of liquid foods. *Journal of Food Engineering*, 49, 201-206.

[28] Malacrida, CR; Motta, S. (2005). Compostos fenólicos totais e antocianinas em suco de uva. *Ciência e Tecnologia de Alimentos*, 25, 659-664.

[29] Muratore, G; Licciardello, F; Restuccia, C; Puglisi, ML; Giudici, P. (2006). Role of Different Factors Affecting the Formation of 5-Hydroxymethyl-2-furancarboxaldehyde in Heated Grape Must. *Journal of Agricultural and Food Chemistry*, 54, 860-863.

[30] Salunkhe, DK; Desai, BB. Postharvest Biotechnology of Fruits. vol. 1, Nova York: CRC Press Inc. 1984.

[31] Grabowski, S; Marcotte, M; Ramaswamy, H. Drying of fruits, vegetables, and spices. In: Ramaswamy, HS; Raghavan, V; Chakraverty, A; Mujumdar, AS. *Handbook of Postharvest Techonology*. Nova York: CRC Press, 2003, 653-995.

[32] Garcia-Parrilla, MC; Leon Camacho, M; Herdia, FJ; Troncoso, AM. (1994). Separation and identification of phenolic acids in wine vinegars by HPLC. *Food Chemistry,* 50, 313-315.

[33] Anklam, E; Lipp, M; Radovic, B; Chiavaro, E; Palla, G. (1998). Characterisation of Italian vinegar by pyrolysis-mass spectrometry and a sensor technique. *Food Chemistry*, 61, 243-248.

[34] Pittia, P; Mastrocola, D; Maltini, E. Evolution of Some Physical Properties of "Aceto Balsamico Tradizionale Di Reggio Emilia"; during Long-Term Aging. In: Buera, MP; Welti-Chanes, J; Lilford, PJ; Corti, HR. *Water Properties of Food, Pharmaceutical, and Biological Materials,* 2006, 671-979.

[35] Falcone, PM; Verzelloni, E; Tagliazucchi, D; Giudici, P. (2008). A rheological approach to the quantitative assessment of traditional balsamic vinegar quality. *Journal of Food Engineering*, 86, 433-443.

[36] Falcone, PM; Giudici, P. (2010). Sugar conversion induced by the application of heat to grape must. *Journal of Agricultural and Food Chemistry,* 58, 8680-8691.

[37] Ramaswamy, H; Meng, Y. Commercial canning of berries. In: Zhao, Y. Berry Fruit. Nova York: CRC Press, 2007.

[38] Gögüs, F; Bozkurt, H; Eren, S. (1998). Kinetics of Maillard reactions between the major sugars and amino acids of boiled grape juice. *Lebensm.-Wiss*, 31, 196–200.

[39] Garde-Cerdán, T; Arias-Gil, M; Marsellés-Fontanet, AR; Ancín-Azpilicueta, C; Martín-Belloso, O. (2007). Effects of thermal and non-thermal processing treatments on fatty acids and free amino acids of grape juice. *Food Control*, 18, 473–479.

[40] Rizzon, LA; Meneguzzo, J. *Suco de uva.* 1ed. Brasília: Embrapa Informação tecnológica; 2007.

[41] Marzarotto, V. Suco de uva. In: Venturini, WGF, editor. *Tecnologia de Bebidas: matéria-prima, processamento, BPF/APPCC, legislação e mercado.* São Paulo: Edgard Blücher; 2005; 311-346.

[42] Pederson, C.S. Grape Juice. In: Tressler, DK; Joslyn, MA. *Fruit and vegetable juice processing technology*. Westport: AVI, 1980; 269-309.

[43] Rizzon, LA; Manfroi, V; Meneguzzo, J. *Elaboração de suco de uva na propriedade vitícola.* Bento Gonçalves: Embrapa Uva e Vinho, 1997.

[44] Rizzon, LA; Link, M. Composição do suco de uva caseiro de diferentes cultivares. *Ciência Rural*, 2006, 36, 689-692.

[45] Nagy, S; Attaway, JA; Rhodes, ME. *Adulteration of fruit juice beverages*. New York: Marcel Dekker, 1988.

[46] Mato, I; Suárez-Luque, S; Huidobro, JF. (2005) A review of the analytical methods to determine organic acids in grape juices and wines. *Food Research International*, 38, 1175–1188.

[47] Rizzon, L.A. & Sganzerla, V.M.A. (2007). Ácidos tartárico e málico no mosto de uva em Bento Gonçalves-RS. *Ciência Rural*, 37, 911-914.

[48] Mietton-Peuchot, M; Milisic, V; Noilet, P. (2002). Grape must concentration by using reverse osmosis. Comparison with chaptalization. *Desalination*, 148, 125-129.

[49] Rizzon, L.A. & Miele, A. (2005). Correção do mosto da uva Isabel com diferentes produtos na Serra Gaúcha. *Ciência Rural*, 35, 450-454.

[50] Rektor, A; Vatai, G; Békássy-Molnár, E. (2006). Multi-step membrane processes for the concentration of grape juice. *Desalination*, 1, 446–453.

[51] Gollücke, APB; Souza, JC; Tavares, DQ. (2008). Sensory stability of Concord and Isabel concentrated grape juices during storage. *Journal of Sensory Studies*, 23, 340–353.

[52] Manach, C; Scalbert, A; Morand, C; Rémésy, C; Jiménez, L.(2004). Polyphenols: food sources and bioavailability. *American Journal of Clinical Nutrition*, 79, 727– 47.

[53] Naczk, M. & Shahidi, F. (2004). Extraction and analysis of phenolics in food. *Journal of Chromatography A*, 1054, 95–111.

[54] Chen, S; Sun, XZ; Kao, Y; Kwon, A; Zhou, D; Eng, E. (1998). Suppression of breast cancer cell growth with grape juice. *Pharmaceutical Biology*, 36, 53–61.

[55] Cheynier, V. (2005). Polyphenols in foods are more complex than often thought. *American Journal of Clinical Nutrition*, 81, 223S–9S.

[56] Joslyn, MA; Farley, HB; Reed, HM. (1992). Effect of temperature and time of heating on extraction of color from red-juice grapes. *Industrial and Engineering Chemistry*, 21, 1135-1137.

[57] Makris, DP; Kallithrakab, S; Kefalas, P. (2006). Flavonols in grapes, grape products and wines: burden, profile and influential parameters. *Journal of Food Composition and Analysis*, 19, 396–404.

[58] Fuleki, T. & Ricardo-da-Silva, J.M. (2003). Effects of cultivar and processing method on the contents of catechins and procyanidins in grape juice. *Journal of Agricultural and Food Chemistry*, 51, 640−646.

[59] Guerra, C; Barnabé, D. Vinho. In: Venturini, WGF. (Org.). *Tecnologia de Bebidas: matéria-prima, processamento, BPF/APPCC, legislação e mercado*. São Paulo: Edgard Blücher, 2005.

[60] Breuil, AC; Adrian, M; Pirio, N; Meunier, P; Bessis, R; Jeandet, P. (1998). Metabolism of stilbene phytoalexins by *Botrytis cinerea*: 1. Characterization of a resveratrol dehydrodimer. *Tetrahedron Letters*, 39, 537-540.

[61] Zhang, Y; Liu, Y; Wang, T; Li, B; Li, H; Wang, Z; Yang, B. (2006). Resveratrol, a natural ingredient of grape skin: antiarrhythmic efficacy and ionic mechanisms. *Biochemical and Biophysical Research Communications*, 340, 1192–1199.

[62] Sautter, CK; Denardin, S; Alves, AO; Mallmann, CA; Penna, NG; Hecktheuer, LH. (2005). Determinação de resveratrol em sucos de uva no Brasil. *Ciência e Tecnologia de Alimentos*, 25, 437-442.

[63] Gollücke, APB; Catharino, RR; Souza, JC; Eberlin, MN; Tavares, DQ. (2009). Evolution of major phenolic components and radical scavenging activity of grape juices through concentration process and storage. *Food Chemistry*, 112, 868–873.

[64] Markakis, P. Stability of anthocyanins in foods. In: Markakis, P. (Ed.) *Anthocyanins as Food Colors*. New York: Academic Press, 1982.

[65] Malacrida, C.R. & Motta, S.D. (2006). Antocianinas em suco de uva: composição e estabilidade. *B.ceppa*, 24, 59-82.

[66] Lima, VLAG; Pinheiro, IO; Nascimento, MS; Gomes, PB; Guerra, NB. (2006). Identificação de antocianidinas em acerolas do banco ativo de

germoplasma da Universidade Federal Rural de Pernambuco. *Ciência e Tecnologia de Alimentos*, 26, 927-935.

[67] Coultate, TP. *Alimentos: a química de seus componentes*. Porto Alegre: Artmed, 2004.

[68] Bobbio, PA; Bobbio, FO. *Química do processamento de alimentos*. São Paulo: Varela, 2001.

[69] Mccallum, JL; Yang, R; Young, JC; Strommer, JN; Tsao, R. (2007). Improved high performance liquid chromatographic separation of anthocyanin compounds from grapes using a novel mixed-mode ion-exchange reversed-phase column. *Journal of Chromatography A*, 1148, 38–45.

[70] Hrazdina, G; Borzell, AJ; Robinson, WB. (1970). Studies on the stability of the anthocyanidin-3,5-diglucosides. *American Journal of Enology and Viticulture*, 21, 201-204.

[71] Rice, AC. Chemistry of winemaking from native american grape varieties. In: Webb, D. (Ed.) *Chemistry of Winemaking*. Dallas: American Chemical Society, 1974.

[72] Brouillard, R; Chassaing, S; Fougerousse, A. Why are grape/fresh wine anthocyanins so simple and why is it that red wine color lasts so long? *Phytochemistry*, 2003, 64, 1179–1186.

[73] Francis, FJ. Analysis of anthocyanins. In: Markakis, P. (Ed.) *Anthocyanins as food colors*. New York: Academic Press, 1982.

[74] Morris, JR; Sistrunk, WA; Junek, J; Sims, CA. (1986). Effects of fruit maturity, juice storage, and juice extraction temperature on quality of 'Concord' grape juice. *Journal of the American Society for Horticultural Science*, 111, 742-746.

[75] Bridle, P. & Timberlake, C.F. (1997). Anthocyanins as natural food colours-selected aspects. *Food Chemistry*, 58, 103-109.

[76] Dufour, C. & Sauvaitre, I. (2000). Interactions between anthocyanins and aroma substances in a model system. effect on the flavor of grape-derived beverages. *Journal of Agricultural and Food Chemistry,* 48, 1784−1788.

[77] Araújo, JMA. *Química de alimentos: teoria e prática*. Viçosa: UFV, 2004.

[78] Galvano, F; La Fauci, L; Lazzarino, G; Fogliano, V; Ritieni, A; Ciappellano, S; Battistini, NC; Tavazzi, B; Galvano, G. (2004). Cyanidins: metabolism and biological properties. *Journal of Nutritional Biochemistry*, 15, 2–11.

[79] Brouillard, R. Chemical structure of anthocyanins. In: Markakis, P. (Ed.) *Anthocyanins as Food Colors*. New York: Academic Press, 1982.

[80] Renaud, S. & De Lorgeril, M. (1992). Wine, alcohol, platelets, and the French paradox for coronary heart disease. *Lancet*, 339, 1523–1526.

[81] Gaziano, JM; Buring, JE; Breslow, JL; Goldhaber, SU; Rosner, B; Van Denburgh, M; Willet, W; Hennekens, CH. (1993). Moderate alcohol intake, increased levels of high-density lipoprotein and its subfractions, and decreased risk of myocardial infarction. *The New England Journal of Medicine*, 329, 1829–1834.

[82] Rajaram, S. (2003). The effect of vegetarian diet, plant foods, and phytochemicals on hemostasis and thrombosis. *American Journal of Clinical Nutrition*, 78, 552S–8S.

[83] Vinson, JA; Teufel, K; Wu, N. (2001). Red wine, dealcoholized red wine, and especially grape juice, inhibit atherosclerosis in a hamster model. *Atherosclerosis*, 156, 67–72.

[84] Williams, MJ; Sutherland, WH; Whelan, AP. (2004). Acute effects of drinking red and white wines on circulating levels of inflammation-sensitive molecules in men with coronary artery disease. *Metabolism*, 53, 318-23.

[85] Coimbra, SR; Lage, SH; Brandizzi, L; Yoshida, V; Da Luz, PL. (2005). The action of red wine and purple grape juice on vascular reactivity is independent of plasma lipids in hypercholesterolemic patients. *Brazilian Journal of Medical and Biological Research*, 38, 1339-1347.

[86] Hansel, B; Thomas, F; Pannier, B; Bean, K; Kontush, A; Chapman, MJ; Guize, L; Bruckert, E. (2010). Relationship between alcohol intake, health and social status and cardiovascular risk factors in the urban Paris-Ile-De-France Cohort: is the cardioprotective action of alcohol a myth? *European Journal of Clinical Nutrition*, 64, 561-568.

[87] Kelly, Y; Sacker, A; Gray, R; Kelly, J; Wolke, D; Quigley, MA. (2009). Light drinking in pregnancy, a risk for behavioural problems and cognitive deficits at 3 years of age? *International Journal of Epidemiology*, 38, 129–140.

[88] Ohno, M; Ka, T; Inokuchi, T; Moriwaki, Y; Yamamoto, A; Takahashi, S; Tsutsumi, Z; Tsuzita, J; Yamamoto, T; Nishiguchi, S. (2008). Effects of exercise and grape juice ingestion in combination on plasma concentrations of purine bases and uridine. *Clinica Chimica Acta*, 388, 167–172.

[89] O'Byrne, DJ; Devaraj, S; Grundy, SM; Jialal, I. (2002). Comparison of the antioxidant effects of Concord grape juice flavonoids and α-

tocopherol on markers of oxidative stress in healthy adults. *American Journal of Clinical Nutrition*, 76, 1367–74.

[90] Bitsch, R; Netzel, M; Frank, T; Strass, G; Bitsch, I. (2004). Bioavailability and biokinetics of anthocyanins from red grape juice and red wine. *Journal of Biomedicine and Biotechnology*, 5, 293–298.

[91] Castilla, P; Echarri, R; Dávalos, A; Cerrato, F; Ortega, H; Teruel, JL; Lucas, MF; Gómez-Coronado, D; Ortuño, J; Lasunción, MA. (2006). Concentrated red grape juice exerts antioxidant, hypolipidemic, and antiinflammatory effects in both hemodialysis patients and healthy subjects. *American Journal of Clinical Nutrition*, 84, 252–62.

[92] Freedman, JE; Parker, C; Li, L.; Perlman, JA; Frei, B; Ivanov, V; Deak, LR; Iafrati, M.D.; Folts, JD. (2001). Select flavonoids and whole juice from purple grapes inhibit platelet function and enhance nitric oxide release. *Circulation*, 103, 2792-2798.

[93] García-Alonso, J; Ros, G; Vidal-Guevara, ML; Periago, MJ. (2006). Acute intake of phenolic-rich juice improves antioxidant status in healthy subjects. *Nutrition Research*, 26, 330– 339.

[94] Stein, JH; Keevil, JG; Wiebe, DA; Aeschlimann, S; Folts, JD. (1999). Purple grape juice improves endothelial function and reduces the susceptibility of LDL cholesterol to oxidation in patients with coronary artery disease. *Circulation*, 100, 1050-1055.

[95] Keevil, JG; Osman, HE; Reed, JD; Folts, JD. (2000).Grape juice, but not orange juice or grapefruit juice, inhibits human platelet aggregation. *The Journal of Nutrition*, 130, 53–56.

[96] Giehl, MR; Dal Bosco, SM; Laflor, CM; Weber, B. Eficácia dos flavonóides da uva, vinho tinto e suco de uva tinto na prevenção e no tratamento secundário da aterosclerose. *Scientia Medica*, 2007, 17, 145-155.

[97] Soobrattee, MA; Neergheen, VS; Luximon-Ramma, A; Aruoma, OI; Bahorun, T. (2005). Phenolics as potential antioxidant therapeutic agents: mechanism and actions. *Mutation Research,* 579, 200–213.

[98] Vita, JA. (2005). Polyphenols and cardiovascular disease: effects on endothelial and platelet function. *American Journal of Clinical Nutrition*, 81, 292S–7S.

[99] Anselm, E; Chataigneau, M; Ndiaye, M; Chataigneau, T; Schini-Kerth, VB. (2007). Grape juice causes endothelium-dependent relaxation via a redox-sensitive Src- and Akt-dependent activation of eNOS. *Cardiovascular Research*, 73, 404–413.

[100] Shukitt-Hale, B; Carey, A; Simon, L; Mark, DA; Joseph, JA. (2006). Effects of Concord grape juice on cognitive and motor deficits in aging. *Nutrition*, 22, 295–302.

[101] Paulo, L; Oleastro, M; Gallardo, E; Queiroz, JA; Domingues, F. (2011). Anti-*Helicobacter pylori* and urease inhibitory activities of resveratrol and red wine. *Food Research International*, 44, 964-969.

[102] Shanmuganayagam, D; Warner, TF; Krueger, CG; Reed, JD; Folts, JD. (2007). Concord grape juice attenuates platelet aggregation, serum cholesterol and development of atheroma in hypercholesterolemic rabbits. *Atherosclerosis*, 190, 135–142.

[103] Partfitt, J; Barthel, M; Macnaughton, S. (2010) Food waste within food supply chains: quantification and potential for change to 2050. *Phil. Trans. R. Soc. B*, 365, 3065-3081.

[104] Lundqvist J; de Fraiture, C; Molden, D. (2008). Saving water: from field to fork—curbing losses and wastage in the food chain. *In SIWI Policy Brief. Stockholm, Sweden: SIWI.*

[105] Grolleaud, M. (2002). Post-harvest losses: discovering the full story. Overview of the phenomen on of losses during the post-harvest system. Rome, Italy: FAO, Agro Industries and Post-Harvest Management Service.

[106] [1]Kosseva, M. R. (2009). Processing of food wastes. *Advances in Food and Nutrition Research*, 58, 57-136.

[107] Kamerer, D, Claus, A; Carle, R; Schieber, A. (2004). Polyphenol screening of pomace from red and white grape varieties (*Vitis vinifera* L) by HPLC-DAD-MS/MS. *Journal of Agricultural and Food Chemistry,* 52, 4360 – 4367.

[108] Silva, L. M. L. R. (2003). Caracterização dos subprodutos da vinificação. Millenium online: Revista do ISPV [ISSN (edição electrónica) -1647-662X], 2003, 28, 123-133. Avaliable from: *http://www.ipv.pt/millenium/ Millenium28/10.pdf.*

[109] Schieber, A; Stintzing, FC; Carle, R. (2001). By-products of plant food processing as a source of functional compounds— recent developments (Review). *Trends in Food Science & Technology,* 12, 401–413.

[110] Crespo, J. & Brazinha, C. (2010) Membrane processing: Natural antioxidants from winemaking by-products. *Filtration and Separation*, 2010, 47(2), 32-35.

[111] Bertran, E; Sort, X; Soliva, M; Trillas, I. (2004). Composting winery waste: sludges and grape stalks. *Bioresource and Technology,* 95, 203-208.

[112] Pérez-Serradilla, J. A. &Luque-de-Castro, M. D. (2011). Microwave-assisted extraction of phenolic compounds from wine lees and spray-drying of the extract. *Food Chemistry,* 124, 1652-1659.

[113] Bustamante, MA; Moral, R; Paredes, C; Pérez-Espinosa, A; Moreno-Caselles, J; Pérez-Murcia, MD. (2008). Agrochemical characterization of the solid by-products and residues from the winery and distillery industry. *Waste Management,* 28, 372–380.

[114] Airoldi, G; Balsari, P; Gioelli, F. (2004). Results of a survey carried out in Piedmont region winery on slurry characteristics and disposal methods. *Proceedings of the 3rd International Specialised Conference on Sustainable Viticulture and Winery Wastes Management Barcelona*, 24–26 May 2004, 335–338.

[115] Shrikhande, A. J. (2000). Wine by-products with health benefits. *Food Research International*, 33, 469–474.

[116] Ferer, J; Páez, G; Mármol, Z; Ramones, E; Chandler, C; Marín, M; Ferrer, A. (2001). Agronomic use of biotechnologically processed grape wastes. *Bioresource Technology,* 76, 39-44.

[117] Bustamante, MA; Pérez-Murcia, MD; Paredes, C; Moral, R; Pérez-Espinosa, A; Moreno-Caselles, J. (2007). Short-term carbon and nitrogen mineralisation in soil amended with winery and distillery organic wastes. *Bioresource Technology,* 98, 3269–3277.

[118] Spanghero, M; Salem, AZM; Robinson, PH. (2009). Chemical composition, including secondary metabolites, and rumen fermentability of seeds and pulp of Californian (USA) and Italian grape pomaces. *Animal Feed Science and Technology*, 152, 243-255.

[119] Niezen, JH; Waghorn, TS; Charleston, WAG; Wagnhorn, GC. (1995). Growth and gastrointestinal nematode parasitism in lambs grazing either lucerne (*Medicago sativa*) or sulla (Hedysarumc oronarium) which contains condensed tannins. *J. Agric. Sci,* 125, 281–289.

[120] Stheinshamn, H; Purup, S; Thuen, E; Hansen-Moller, J. (2008). Effects of Clover-Grass Silages and Concentrate Supplementation on the Content of Phytoestrogens in Dairy Cow Milk. *Journal of Dairy Science,* 91, 2715-2725.

[121] Valko, M; Rhodes, CJ; Moncol, J; Izakovic, M; Mazur, M. (2006). Free radical, metals and antioxidants in oxidative stress-induced cancer. *Chemico-Biological Interactions,* 160, 1-40.

[122] Scandalio, J. G. (2005). Oxidtive stress: molecular perception and transduction of signals triggering antioxidant gene defense (review). *Brazilian Journal of Medicine and Biological Research,* 38, 995-1014.

[123] Middleton Jr., E; Kandaswami, C; Theoharides, TC. (2000). The effects of plant flavonoids on mammalian cells: implications for inflammation, heart disease and cancer. *Pharmacol. Rev.,* 52, 673-839.

[124] Montealegre, RR; Peces, RR; Vozmediano, JLC; Gascuena, JM; Romero, EG. (2006). Phenolic compounds in skins and seeds of ten grape Vitis vinifera varieties grown in a warm climate. *Journal of Food Composition and Analysis*, 19, 687–693.

[125] Rockenbach, II; Gonzaga, LV; Rizielo, VM; Gonçalves, AESS; Genovese, MI; Fett, R. (2011). Phenolic compounds and antioxidant activity of seed and skin extracts of red grape(*Vitis vinifera* and *Vitis labrusca*) pomace from Brazilian winemaking. *Food Research International,* 44, 897-901.

[126] Bail, S; Stuebiger, G; Krist, S; Unterweger, H; Buchnauer, G. (2008). Characterization of various grape seed oils by volatile compounds, triacylglycerol composition, total phenols and antioxidant capacity. *Food Chemistry*, 108, 1122- 1132.

[127] Rockenbach, II; Rodrigues, E; Gonzaga, LV; Fatt, R. (2010). Fatty acid composition of grape (*Vitis vinifera* L. and *Vitis labrusca* L.) seed oil. *Brazilian Journal of Food Technology*, III SSA, 23 -26.

[128] Nawaz, H; Shi, J; Mittal, GS; Kakuda, Y. (2006). Extraction of polyphenols from grape seeds and concentration by ultrafiltration. *Separation and Purification Technology,* 48, 176-181.

[129] Wada, M; Kido, H; Ohyama, K; Ichibangase, T; Kishikawa, N; Ohba, Y; Nakashima, MN; Kuroda N; Nakashima, K. (2007). Chemiluminescent screening of quenching effects of natural colorants against reactive oxygen species: Evalutation of grape seed, monascus, gardenia and red radish extracts as multi-funcional food additivies. *Food Chemistry,* 101, 980-986.

[130] Gómez-Plaza, E; Miñano, A; López-Roca, JM. (2006). Comparison of chromatic properties, stability. *Food Chemistry,* 97, 87-94.

[131] Lapornik, B; Prosek, M; Wondra, AG. (2005). Comparison of extracts prepared from plant by-products using different solvents and extraction time. Journal of Food Engineering, 71, 214–222.

[132] Corralles, M; Toepfl, S; Knorr, D; Tauscher, B. (2008). Extraction of anthocyanins from grape by-products assisted by ultrason high hydrostatic pressure or pulsed electric fields: A comparison. *Innovative Food Science and Emerging Technologies,* 9, 85–91.

[133] Saura-Calixto, F. (1998). Antioxidant dietary fiber product: a new concept and a potential food ingredient. *Journal of Agricultural and Food Chemistry,* 48, 4303-4306.

[134] Llobera, A. & Cañellas, J. (2007). Dietary fibre content and antioxidant activity of Manto Negro red grape (*Vitis vinifera*): pomace and stem. *Food Chemistry,* 101, 659-666.

[135] Sánchez-Alonso, I; Jiménez-Escrig, A; Saura-Calixto, F; Borderías, A J. (2007). Effect of grape antioxidant dietary fibre on the prevention of lipid oxidation in minced fish: Evaluation by different methodologies. *Food Chemistry,* 101, 372-378.

[136] Botella, C; Diaz, A; Ory, I; Webb, C; Blandino, A. (2007). Xylanase and pectinase production by *Aspergillus awamori* on grape pomace in solid state fermentation. *Process Biochemistry,* 42, 98-101.

[137] Parathaman, R; Vidyalakshmi, R; Murugesh, S; Singaravadivel, K. (2009). Effects of Fungal Co-Culture for the Biosynthesis of Tannase and Gallic Acid from Grape Wastes under Solid State Fermentation, *Global Journal of Biotechnology and Biochemistry,* 4, 29-36.

[138] Silveira, ST; Daroit, DJ; Brandelli, A. (2008) Pigment production by Monascus purpureus in grape waste using factorial design. *LWT Food Science and Technology,* 41, 170–174.

[139] Bracchitta, M; Zanichelli, D; Setti, L. (2010). Special Abstracts in: Journal of Biotechnology, 150S, S1–S576.

[140] Xu, R; Ferrante, L; Briens, C; Berruti, F. (2009). Flash pyrolysis of grape residues into biofuel in a bubbling fluid bed. *Journal of Analytical and Applied Pyrolysis,* 86, 58-65.

[141] Ping, L; Brosseb,N; Sannigrahi, P; Ragauskac, A. (2011). Evaluation of grape stalks as bioresource. *Industrial Crops and Products,* 33, 200-204.

[142] Dinuccio, E; Balsari, P; Gioelli, F; Menardo, S. (2010). Evaluation of the biogas productivity potential of some Italian agro-industrial biomasses. *Bioresearch Technology,* 101, 3780-3783.

[143] González-Centeno, MR; Rosselló, C; Simal, S; Garau, MC; López, F; Femenia, A. (2010). Physico-chemical properties of cell wall materials obtained from ten grape varieties and their by-products: grape pomaces and stems. *LTW – Food Science and Technology,* 43, 1580-1586.

[144] Fiol, N; Ecuerdo, C; Vilaescusa, I. (2008). Chromium sorption and Cr(VI) reduction to Cr(III)by grape stalks and yohimbe bark. *Bioresource Technology,* 99, 5030-5036.

[145] Deiana, AC; Sardella, MF; Silva, H; Amayab, A; Tancredi, N. (2009). Use of grape stalk, a waste of the viticulture industry, to obtain activated carbon. *Journal of Hazardous Materials,* 172, 13-19.

[146] Cayuela, ML; Sánchez-Monedero, MA; Roig, A. (2010). Two-phase olive mil waste composting: enhancement of the composting rate and compost quality by grape stalks. *Biodegradation,* 21, 465-473.

[147] Spigno, G. & Faveri, D. M. (2007). Antioxidants from grape stalks and marc: Influence of extraction procedure on yield, purity and antioxidant power of the extracts. *Journal of Food Engineering,* 78, 793–801.

[148] Cárcel, JA; Garcia-Pérez, JV; Mulet, A; Rodríguez, L; Riera, E. (2010). Ultrasonically assisted antioxidant extraction from grape stalks and olive leaves. *Physics Procedia,* 3, 147-152.

[149] Garcia- Perez, JV; Garcia-Avarado, MA; Carcel, JA; Mulet, A. (2010). Extraction kinetics modeling of antioxidants from grape stalk (*Vitis vinifera* var. Bobal): Influence of drying conditions. *Journal of Food Engineering,* 101, 49-58.

[150] Salgado, JM; Rodríguez, N; Cortés, S; Domínguez, JM. (2010). Improving downstream process to recover tartaric acid, tartrate and nutrients form vinasse and formulation of inexpensive fermentative broths for xylitol production. *J. Sci. Food Agric.,* 2168-2177.

INDEX

A

B

C

D

G

H

I

N

O

P

Q

R

S

T

U

V

W

X

Y

Z